*Whence thine eyes see equanimity
in all things…this is peace.*

WHAT IS GOD?

ROLLING BACK THE VEIL

CHRISTINE HORNER

in the garden Publishing

a media company of
WHAT WOULD LOVE DO INT'L LTD

ISBN: 978-0-9888333-3-3 Paperback
ISBN: 978-0-9888333-2-6 Hardcover
ISBN: 978-0-9888333-4-0 e-Book

Library of Congress data available upon request.

Published by:

IN THE GARDEN PUBLISHING
P.O. Box 752252
Dayton, OH 45475

www.IntheGardenPublishing.com
www.WhatWouldLoveDoIntl.com

CONTENTS

For My Children, A & V,

It is because of you that I began this journey, seeking a better life, only to discover I already had it.

My deepest apologies for dragging you along with me.

I hope that someday you can see the view from where I am.

INTRODUCTION

The human race is exhausted.

Human behavior has reached the height of ridiculousness. The only thing left seems unfathomable. Yet the abyss looms large and tangible enough in the corner of our eye to keep us moving away from it, toward something... anything that is better than *this*.

Look out your window and see that there is very little in the world that has been created by man that is of practical value. In spite of documents signed and millennium goals set, we are still mired in age-old constructs that perpetuate conflict and human suffering.

Whether you are gazing out into your own backyard or the world at large, it seems that much of humanity is still gridlocked in a state of survivalism and competition for access to basic necessities such as food, water, shelter, healthcare and education—even in the industrialized nations.

When we expand our gaze, we see that the Earth is abundant enough to feed, house and even nurture a population of this size. We see our own capacity to mirror nature through working together cooperatively to create a renewable and sustainable existence, in harmony with all of Life around us that is inclusionary rather than exclusionary.

Further, something does appear to be happening. In addition to the announcement of one of the most important scientific discoveries in recent history, human consciousness is undergoing a shift so profound, the proof can be seen with the naked eye.

The intent of this book is not to bring you hope. Hope is the last thing humanity needs at this, "the hour of our death." This book is here to dash every hope you've ever had so that you may finally be able to move beyond hope into something more substantial, something more powerful; NOW, though no more physical.

With half the world still mired in war and poverty and the "good" half amidst economic chaos, deteriorating political and social infrastructures, you may find yourself on your knees, hands folded in prayer with nothing left to say, the silence deafening as you strain for a sign—any sign.

Perhaps in a collapse of faith, you find yourself looking skyward with tears in your eyes, hurt and confused. How could God continue to allow such suffering in the world? Why aren't things getting better?

Or, you might be part of the once mirthful camp, now more somber than gleeful, that thinks this is proof enough that God does not exist at all.

Even world-renowned Physicist, Dr. Stephen Hawking, ditched his prior position in his book, *The Grand Design*: "belief in a creator was not incompatible with science" and now concludes that the Big Bang was as inevitable as gravity.

Dr. Hawking goes on to say, "Spontaneous creation is the reason there is something rather than nothing, why the universe exists, why we exist. It is not necessary to invoke God to light the blue touch paper and set the universe going."

For those of you sitting in full Samadhi in a state of bliss, by all means, as you were; you've already got this book figured out.

For the rest of us, you no doubt have begun to ask questions — perhaps for the first time in your life. If you were raised under the flag that blind faith rather than inquiry is the path to Heaven, *now* is the time to ask yourself, "Can I absolutely be certain of that?"

For this is no ordinary inquiry. This inquiry will turn you upside down and inside out. You will be asked to question everything you hold dear and think you know to be true — even who you are relative to the world. Even...GOD.

PART I

CHAPTER 1

Who Am I?

Let's sojourn together. Are you willing to explore in a way you may never have done before?

This will require no mode of transportation and you will not go anywhere, yet it may be so exhausting, in the words of one sojourner named Tom, "I couldn't get out of bed for three weeks when I finally 'got it.'"

Close your eyes. Look *within* your body. Notice its steady, rhythmic breathing occurring without your needing to monitor it. Now, allow your mind's eye to notice and feel the rest of your body.

Allowing yourself to relax and sink more deeply toward the Earth, you feel the edges of your body, of where your body ends and your chair or bed begins—yet you feel *both* your back and hips and the chair at the same time.

As you continue to let go of the tension (control), the line between "out there" and "in here" begins to blur. Have you ever been so still that you no longer feel your body?

Are you ready? You are safe.

Your body has relaxed fully and is connected with and grounded to the Earth. Now, allow your mind's eye to expand its awareness outside your body. It's okay. Don't be afraid.

In wonder and awe, all thoughts have ceased. You become Awareness Itself which is beyond the human part of you. You/ Awareness expand outside the room you are in, the building or vehicle... Your body has disappeared. Breathing is no long required. The emotional anxiety and even pain you've carried with you and have been running from every moment of your existence has vanished. You may not even have been conscious of it until now. Free of its weight, you experience a lightness you've never known possible.

You/Awareness widen further, as if you are pouring *into* your surroundings, yet there is no sky above, there is only that of which you are aware. Things once perceived as separate from you, you now **are.** The cars in front of you, the buildings next to you, the stretches of forests and farmland, the deserts, the mountains and the oceans. Everything in the world has taken on a transparent quality and gone silent. Keep going.

Awareness (you) has expanded so vastly that you comprehend that you are the Earth itself; not only the Earth, but the wind, the sky and the space that surrounds it. The moon seems to come 'round to greet you. Giddy with delight, a knowing that you could keep going fills you, and in the space surrounding you you become the nebulae and clusters of stars and galaxies themselves just by turning awareness toward it—yet you understand that there is nowhere to go because you are now-here. *All contained within.*

A peace so utterly still fills you, a vacuum is created allowing joy like you've never known to overtake you. So pure is this joy, that it is orgasmic in nature. The desire to laugh hysterically and weep like you've never wept before rushes through you all at once until you center once again on the all pervasive peace and stillness of all that is in this present moment.

Remaining in this timeless vantage point in space, you turn your gaze from the stunning moon to beautiful planet Earth. A single, indivisible cerulean blue ball that is part of you, so exquisite, an overwhelming sense of love fills you, the boundary between you and home obliterated.

Ask yourself…

WHO

AM

I?

CHAPTER 2

The Value of Inquiry

Thoughts begin to return. Thoughts create the appearance of time and space. Thoughts are separation itself as a condition of relativity. Thoughts are parts of Creation's self-sustainable, organizational structure...

It is human nature to look out into the world for points of reference to gather data. As your eyes survey what lies "out there," the wondrous miracle of the biological instrument called your brain, very much still a mystery in itself, sets to work.

The region of your brain called the pre-frontal cortex compares sensory stimulation against its own unique catalog, your memory, relational to the personal "self" in order to generate context; sort of a mental sketch pad.

Chapter 2: The Value of Inquiry

The typical psychological term for the functions carried out by the pre-frontal cortex is called executive function. Executive function has been connected to planning complex cognitive behavior, personality expression, decision making and moderating social behavior.

As the pre-frontal cortex occupies a far larger percentage of the brain in humans than in any other animal, it is widely believed that it relates directly to sentience; the ability to have subjective experiences. Context relative to self is important, in the final analysis, for the purpose of recognition; **self-recognition** to be exact, or consciousness.

Sentience is necessary for the ability to suffer. Eastern religions recognize non-humans as sentient beings, though sentience is then placed on a scale of one to five. The first degree of sentience is touch, with the addition of degrees by order of the senses until you arrive at humans, a sentient being of the fifth order. In Buddhism, there is recognition of a sixth sense; the subjective experience of the mind.

Consciousness is what makes human experience possible.

Subjective experiences are strung together forming what appears as a linear timeline that represents the past, the present and even the ability to imagine a future. Internal comment- ary or dialogue referred to as thoughts create the context for forming personalized judgments. Thus, you now have created up and down, near and far, light and dark, right and wrong, good and bad, pleasure and pain. This is useful for creating **self-organization**.

This very basic introduction into a few of your complex brain operations is to illustrate your brain's function and purpose on a local level in this 3D contextual playing field of relativity called the Universe.

Not just through touch, but also through the four other tools of perception thought to be part of the body; sight, hearing, taste and smell, we experience all that life has to offer.

However, this is just the beginning. Perception is hardly limited to only the five senses and not to just the body. Unlike much of the West, other cultures are acutely aware of this knowledge which is deeply embedded within their heritage and daily life as they seek guidance from the realms of the more mysterious.

To their credit, the global scientific community has been exploring the more "mysterious" as they attempt to prove the existence of the "God particle" called the Higgs boson. The Higgs boson particle is thought to be the point of origin at which particles gain mass.

With the July 4th, 2012 announcement via the multi-national, underground Large Hadron Collider, and whether the scientific and religious communities like it or not, we can safely say we've bridged the centuries-held divide between the two communities that literally brought death and destruction throughout the various inquisitions as political and religious leaders, grossly interconnected, fought to hold onto power they felt the scientific community could threaten to take away.

That or it's a great marketing ploy.

Through inquiry, there is discovery. The most famous discovery of all made in the allegory of Adam and Eve in the Garden of Eden, when Eve was tempted by a snake, a symbol of evil, to eat the apple from the tree of Knowledge.

The Adam and Eve story is incomplete as told and has even been manipulated in the re-telling, yet it exists in most cultures in some form.

Have you ever truly inquired into the nature of the Garden of Eden and Adam and Eve? Even if you give it more than two minutes contemplation, you would begin to have questions. For example, how could evil have entered God's perfect Garden of Eden to begin with? Why is Eve the scapegoat for the fall of Man as a whole?

Thousands of years into recorded human history, though we've advanced technologically as a species, very little has been resolved in regard to our existential dilemmas; even with the aid of sages, scientists, philosophers, charities, world leaders, a savior, humanitarians, and, yes... GOD.

Repeating the same questions, "why is there so much evil in the world?", and "why isn't God doing anything about it?" has not really changed anything in the last several centuries. Even if you don't believe in God and repeat your own version of the questions, clearly something is amiss. This is why inquiry is imperative now more than ever.

It is through personal inquiry that the impersonal is revealed. As the Source of All, by whatever name you prefer, is said to be Absolute and you, by virtue of death, are not, perhaps it's time we change the questions. Consider that the world we see is but a symptom of a causeless cause. To understand our external world, it is time once again to inquire within.

CHAPTER 3

The Elephant and the
Three Blind Men

It is identification with thoughts that gives rise to self-consciousness or ego, though not a separate entity unto itself. Where do thoughts come from? Do thoughts of loving adoration and devotion toward your family or even a piece of chocolate cake come from one place and thoughts of revenge or suicide come from another? What thoughts are actually original versus those that mirror collective culture and conditioning?

It is the nature of the mind to corrupt truth. Thoughts filter through the pre-frontal cortex of the brain, which distorts truth via subjective partitioning. Nearly all individual and collective human behavior is based on what amounts to incomplete perceptions formed by the brain based on limited data. The five senses allow us to experience life, but at the same time, they limit the experience we have down to what the brain, as a receiver, can process. Nowhere is this better illustrated than with the Parable of the Elephant and the Blind Men.

Originating in India, but crossing many religious traditions, the parable has been used historically and in modern times to demonstrate the "many-sidedness" of truth and insight into the relativism, the inexpressible nature of Truth and respect for different perspectives.

From the Buddhist Canon Udana 68-69:

"A number of disciples went to the Buddha and said, 'Sir, there are living here in Savatthi many wandering hermits and scholars who indulge in constant dispute, some saying that the world is infinite and eternal and others that it is finite and not eternal, some saying that the soul dies with the body and others that it lives on forever, and so forth. What, Sir, would you say concerning them?'

"The Buddha answered, 'Once upon a time there was a certain raja who called to his servant and said, 'Come, good fellow, go and gather together in one place all the men of Savatthi who were born blind... and show them an elephant.' 'Very good, sire,' replied the servant, and he did as he was told.

"He said to the blind men assembled there, 'Here is an elephant,' and to one man he presented the head of the elephant, to another its ears, to another a tusk, to another the trunk, the foot, back, tail, and tuft of the tail, saying to each one that that was the elephant.

"When the blind men had felt the elephant, the raja went to each of them and said to each, 'Well, blind man, have you seen the elephant? Tell me, what sort of thing is an elephant?'

"Thereupon the men who were presented with the head answered, 'Sire, an elephant is like a pot.' And the men who had observed the ear replied, 'An elephant is like a winnowing basket.' Those who had been presented with a tusk said it was a ploughshare. Those who knew only the trunk said it was a plough; others said the body was a grainery; the foot, a pillar; the back, a mortar; the tail, a pestle, the tuft of the tail, a brush.

"Then they began to quarrel, shouting, 'Yes it is!' 'No, it is not!' 'An elephant is not that!' 'Yes, it's like that!' and so on, till they came to blows over the matter.

"'Brethren, the raja was delighted with the scene.

"'Just so are these preachers and scholars holding various views blind and unseeing... In their ignorance they are by nature quarrelsome, wrangling, and disputatious, each maintaining reality is thus and thus.'

"Then the Exalted One rendered this meaning by uttering this verse of uplift,

O how they cling and wrangle, some who claim

For preacher and monk the honored name!

For, quarreling, each to his view they cling.

Such folk see only one side of a thing. "

The question remains, does many-sidedness

come together to complete truth or ultimately invalidate itself all together as truth can never be fully known by the individual, but rather can only be conceptualized?

The Indian Jain religion version of the story has a king explaining to the blind men:

"All of you are right. The reason every one of you is telling it differently is because each one of you touched a different part of the elephant. So, actually the elephant has all the features you mentioned."

Better yet, 13th Century Persian poet, Rumi, in his retelling, *"Elephant in the Dark"* ends his poem stating, "If each had a candle and they went in together the differences would disappear."

What can be surmised from the parable is that it appears that all perceptions contain *aspects* of truth but do not represent truth itself. No individual or even collective group has the ability to be aware of every facet of any *thing* that exists as either a micro- and/or macrocosm—which in an interconnected Universe, all things exist as both.

Even one claiming to be transcribing truth directly from God above will subject the transcription to the claimant's own inherent limitations as a seer. Rather than arguing over whose truth is more accurate, what if we did as Rumi suggested and focused on the commonality of our discoveries?

Clearly, if the holy writings of various scriptures were each any closer to the truth than the other, it would be abundantly self-evident which one, after a couple of thousand years. We'd be more likely to be taken under its advisement in an effort to eradicate the evil that exists in the world. Yet, those that do follow the teachings don't seem to be overcoming the world and its evil any better than the rest of us. And material wealth at the expense of other humans and the planet itself is not indicative of sophistication or superiority.

Is this an over-simplification of the predicament of man? Yes… and no. It is the nature of the ego to divide for the purpose of conquering—meaning to be under the false belief that we have the ability to control our external reality via external means. Remember, nearly all human behavior is predicated on limited points of view created by thoughts, sensory input and the other biggie—culture and conditioning.

What if dividing life up is the problem itself? What if we "reverse engineer" life and begin to look at it from a more holistic point of view? What if, rather than focusing externally toward others for answers, peace, solutions, happiness, prosperity, leadership, love and fulfillment, we turn inward?

Is it possible to transcend the limitations of the mind and the human experience to come to a higher understanding of the Universe, your role in it, and even of God? How can this knowledge be used to tap into our innate wisdom to end misery in the world, which begins with you?

Until now, there have only been a small number of individuals who have been able to transcend and even ascend the world, even though the seed of possibility exists inside each one of us. The common element between Buddha, Jesus, Krishna and those in modern times such Ramana Maharishi, Papaji and Eckhart Tolle is that they saw the fruitlessness of looking out into the world for the solution to the end of their own personal suffering and the evil in the world.

Enlightened ones such as Gandhi, Martin Luther King, and even John Lennon found their candles within, gently blew on the flame, becoming lights unto the world, thus allowing all of us to see the light that exists within each one of us.

Do we abandon the external world totally? Of course not. You *are* the living mystery made manifest. All of life in the visible and the invisible realms conspires to propel you toward knowing who and what you really are and why you are here. For you to know yourself is the basis of your very existence. There has only ever been one reason; to desire to stay in the dark or to perpetuate mystery.

When the invisible realm seems too intangible or even daunting, each one of us can embrace the mystery of life by observing the external world of temporal form as a starting point to come to also know your inner world of formlessness, which remains changeless and Absolute. *For you are also both.* The formless exists within form within the formless within the form within the formless... This is true and observable even within our own Universe.

Back at our vantage point, nearly 240,000 miles from where you now sit, your home planet Earth operates as a single living organism. Turning your gaze outward, you see what appear to be clusters of stars and colorful clouds of gas that fill you with awe and wonder. If you were to physically traverse the Milky Way galaxy, you'd notice it was mostly space, emptiness, formlessness... Travel further and you would discover the Universe is filled with galaxies with what amounts to not much more than dust for the actual physical contents of it. The Universe, including you, is 99.9% empty space, yet contains all the necessary building blocks for life, including the God Particle of formlessness.

Let's reverse course for just a bit in our sojourn, from the macro to the micro, by turning your focus back toward home. To lay ground for a path for the expansion of your comfort zone when we journey into more difficult subjects, let's first examine what we know about the Universe at large.

Your eyes penetrate the cloud cover, you move closer and closer until you penetrate an atom. An atom is so small at one tenth of a millionth of a millimeter across, if you were to refer to the width of a human hair; it would take a million atoms across to cover the distance.

Inside the atom, surrounded by electrons, we encounter, once again, 99.9% empty space, the weight of an atom being centered in the nucleus. At a trillionth of a millimeter across, if you were to scale the nucleus to the size of a football, the nearest electron would be a half mile away.

The discovery that the atom is mostly space was made by Ernest Rutherford of Manchester

University in 1909. This astounding discovery flew in the face of what was believed by scientists up until then; hence, along with Einstein's Theory of Relativity, we have the birth of quantum mechanics.

If you were to take the space out of the atoms in your body, you would be a lot smaller than a grain of salt. Further, until atoms collapse through conscious observation, they exist in wave form and in more than one place at a time, known as Heisenberg's Uncertainty Principle, now a fundamental concept of Quantum Physics.

Though the Heisenberg Uncertainty Principle has to do with the inability to measure energy at a precise time due to the physical variables intrinsic to a particle's state, more importantly, it reveals the Principle is not about measurement at all but the nature of a particle— its actual fluctuations within an elemental particle as to its energy or momentum.

What is most notable about the contraction of atoms from a wave-like state to a particle is that it is instantaneous rather than evolutionary in nature. *The catalyst appears to be observation.* As science is rooted in materialism, what about the "elephant in the room" that tends to be rejected by materialists: conscious observation?

CHAPTER 4

The Science of Creation

There are two great theories in physics that scientists have spent decades seeking to unify: The theory of the very large; Einstein's Theory of Relativity, and the theory of the very small; Quantum Mechanics. These theories will be discussed more in-depth in the following chapter.

Whether or not you view life from the very large to the very small, there are many clues that can be gathered as to the nature of Creation. This can come from spending time in and around nature itself. One of the first things you notice is life is a metaphor for life is a metaphor for life is a metaphor for life-extending into the infinite.

But why is this so? *Creation by its very nature is infinite possibility. Everything seen and unseen is possible.* What can be found at the micro or local level is also found at the macro or non-local level in self-similar patterns. Examine the spiraling bracts of a pinecone, a tree from leaf to branch to trunk, or the inside of a nautilus shell at any level of its organization and you will see that this is so.

Fig. 1 Nautilus Shell

The patterns are self-similar, not *exactly* the same in that the world of form is *relative*, or relational. Not even human twins are exactly the same. No two snowflakes are exactly alike at the molecular level. Self-similarity is also required in that all of life must be mutually beneficial in order to continue to exist. Every part of Creation is contained *within* Creation in the same way that every cell in your body contains your full DNA blueprint, in that your cells are capable of replicating themselves.

During his youthful travels with his father, Italian mathematician and merchant, Leonardo Pisano Bigollo discovered the Hindu system of numerals from one to nine from the Arabs. In 1209 A.D., Bigollo (Fibonacci to his friends), wrote a book called *Liber Abaci* (*The Book of Computation*) offering an alternative to the clumsy system of Roman numerals and Greek letters, convincing Europe to adopt the numeral system known today as the Hindu-Arabic numeral system.

Leo also introduced to the Western World what has become known as the Fibonacci Series or Phi; though, the Fibonacci numbers derive from Hindu mathematicians and the Arab scholars who preserved them.

The original series is constructed from the numbers, 0 and then 1, and then adding the last two numbers in the series to obtain the next number. For example, 0+1=1, 1+1=2, 1+2=3, 2+3=5, 3+5=8...

The Fibonacci sequence becomes:

0 1 1 2 3 5 8 13 21 34 55 89 144 233 377
610 987 1597 2584 4181 6765 10,946 17,711
28,657 46,368 75,025 121,393 196,418 17,811...

Continuing ad infinitum, the resulting mathematical ratio has become known as the Golden Mean *Spiral,* a self-similar pattern, or iteration, found in Nature from honeybees to plants and animals to the human skeleton, built upon the length of a single dimensional line. Ancient Greeks and Egyptians used the Golden Mean when building structures and monuments, including the Great Pyramid. Plato called this value, "The key for the universe physics."

PHI, 1.618... has no arithmetic solution, continuing into eternity, thus indicating the infinite nature of Creation. Da Vinci studied its mathematical properties extensively, including Fractal Geometry. Artists and architects today still use the Golden Ratio Spiral of 1.618... to create a feeling of "order" in their work. Fig. 2 shows the equation for the Golden Mean Ratio. Even if the mathematical formula is meaningless to you, you can see the fractal and holographic nature of the variables.

(For the sake of brevity, please learn more about how the Golden Mean Ratio is calculated, online.)

$$\frac{a \qquad b}{a+b}$$

$$\frac{b}{a} = \frac{a}{a+b}$$

Golden Section

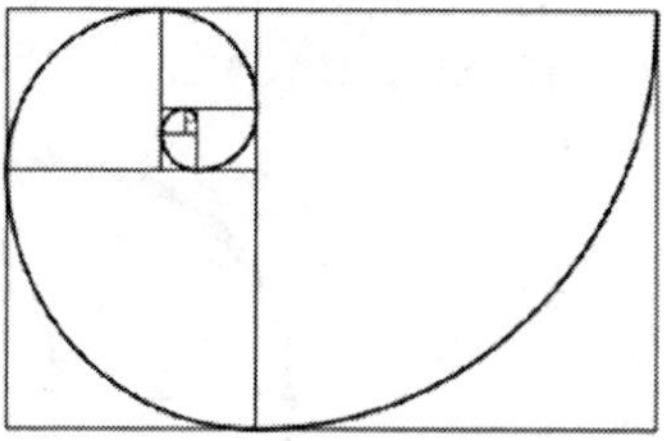

Composition of golden rectangles & Fibonacci spiral

Fig. 2

Let's return to the series and examine the first two numbers; zero, which represents the Absolute or the formless, and one, which represents an energy form, the manifest Monad (unit). The combination of **zero**/formless + **one**/monad yields another monad (energy form). We now have two monads, or complementary energy forms which represent the expansion of unity into duality as the negative/positive aspects of Creation.

The overlapping intersection of the feminine and masculine monads in Fig. 3 form the center, known as the Vesica Pisces, the two dimensional images giving rise to all of Creation as three dimensional form.

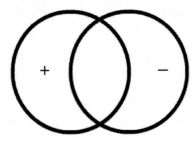

Fig. 3 Overlapping Monads with
Vesica Pisces in Center

Note that the above illustration also resembles the human brain, in that the two distinct hemispheres joined together by the corpus callosum—a series of "cables" that connect all of the information on the left with all of the information on the right—come together to create 3D images.

The Vesica Pisces and its more evolved and complex structures, the Flower of Life and the Tree of Life, have a history of thousands of years, and its symbols can be found embedded in nearly all major religions and in historical buildings and monuments around the globe.

The far deeper purpose in the development of the sequence of numbers, and its spiral appearance, is one which provides a profound image of Creation appearing to move through time or evolve through the resultant fractals.

Numbers are not just linear in nature, but are *relational*, in that they exist at far greater "depth" than we currently understand in base-ten mathematics. Three profoundly influences four, four profoundly influences three and five, and so on.

When Galileo said, "Mathematics is the language from which God has written the Universe," he was referring to the geometry of the Universe which is concerned with the properties, measurement and relationships of structure in space.

Euclidean Geometry, what most of us learned in school, came from the Greek text *The Elements of Euclid*, written around 300 B.C.E. Though Euclidean Geometry has allowed humans to travel to the moon, it falls short in its ability to replicate the complex structures of Life.

Benoit Mandelbrot was a 20th century French-American mathematician best known as the father of fractal geometry through his work presented in *The Fractal Geometry of Nature*. Though he coined the term "fractal," his work had been built upon the work of mathematicians Gaston Julia and Pierre Fatou before him, using what was unavailable to them at the time, a computer.

In 1975, in an IBM computing lab, Mandelbrot was able to solve the millions of iterated functions involved in a formula using a simple equation of addition and multiplication repeated ad infinitum.

This simple equation generated the most stunning organic and complex images ever generated by fractal formulae, as nested self-similar patterns. Within the complexity, patterns in Nature were demonstrated which prior had been called curious coincidences by science. Mandelbrot's fractal geometry clearly documented the relationship between patterns in a whole structure and patterns in its parts.

Creation's fractal nature can be useful toward understanding genetics, DNA, the notion we have "lessons" to learn, and the more esoteric concepts of karma and reincarnation.

Universe Not Fractal?

In August, 2012, it was announced that Morag Scrimgeour at the International Centre for Radio Astronomy Research (ICRAR) at the University of Western Australia in Perth, and her colleagues, using the Anglo-Australian Telescope, were able to probe and survey the Universe at larger scales than any survey before it.

Researchers found that matter is distributed very evenly throughout the universe on extremely large distance scales, with little sign of fractal-like patterns. Scrimgeour was relieved as, "it would mean our whole picture of the universe could be wrong," as cosmologists were modeling the Universe using Einstein's Theory of Relativity.

What's important to note is that, again, researchers are only seeing a part of the whole. Is the Universe regarded as an open or closed system? As Creation is infinite, consider that our Universe is, in itself, a self-similar fractal or part of a continuing system of greater and greater complexity.

Let's use snow as an example on a local level. If the Universe is truly fractal, this pattern and many others you will think of will hold up under inquiry. Snow is made up of millions of snowflakes, no two alike, yet when you put them all together; it appears as a smooth unit. Yet, Creation does not stop at snow. Are we assuming Creation ends at the "corners" of our Universe and the third dimension? What if our Universe is a unique snowflake where Einstein's Theory of Relativity holds up "locally," but not beyond? What of the many other dimensions beyond the four we barely understand?

To attempt to explain our Universe using unsophisticated base-ten mathematics is naive at best. Look at your children or your parents if you don't have children, and ask yourself if the Universe is fractal. Until mathematics beyond base-ten is made part of our educational systems, human perception of the mystery of Creation will continue to be simplistic.

Whether you observe Nature on a local level from your backyard or the intricacy of the cosmology through an Anglo-Australian Telescope, you can easily garner enough sophistication to conclude that holistically Nature works together cooperatively and in harmony, rather than competitively.

Again, we return to the question, why is this so? Structures become shackles. This is guaranteed as any three-dimensional construct contains a positive *and* a negative charge as illustrated in the Yin-Yang symbol below. Note how the white half contains a small circle of black and vice versa.

Fig. 4 Yin-Yang Symbol

In any human construct, both the positive and negative features are evidentially seen to ensure we endeavor to expand our consciousness to seek out higher Truth.

Evolution is Creation in process. Humanity cannot hold onto any experience or belief forever, even if we try. Even conscious awareness expands fractally. In the realm of form, ALL structures are meant to pass away to allow for the appearance of more evolved structures—meaning the appearance of higher Truth, or more succinctly put, more *inclusive* structures of thought as a means to a higher purpose.

Does this mean man and woman should continue to be condemned as fallen from grace, playthings for outside powers such as God and the devil, or is there a much larger point of view to discover?

Perfection lies in all Creation, even that which we do not understand and judge to be imperfect. You cannot change a person, thing, system or society by being part of it. It takes Creation's fractal expansion for evolution to be witnessed. This can be seen on every level of life molecularly, biologically, ecologically, socially, economically...

The logic that our Universe is a unique "snowflake" unto itself can also be applied in that perhaps Creation is not fractal beyond our Universe. It is safe to say that *this* Universe is fractal in design. What is beyond it would necessitate a mutually beneficial self-similar pattern for Creation to thrive. (Still sounds rather fractal.)

Even the realm of the formless is unable to exist without the sustainability of the realm of form. This is known as...

49

CHAPTER 5

The Divine Paradox

Though Creation exists as infinite possibility not all possibility manifests physically. We know this to be true as not every thought you think manifests into your reality. Not every consummative act results in offspring, not every collision of particles results in mass. Not every prayer, desire or bright idea materializes into fruition. And, thankfully so.

As the point of Creation is to thrive and all is interconnected, not every possibility, which cannot exist in detached singularity, is in the highest and best interest of the Whole to manifest beyond thought into form. Infinite possibility exists as physical manifestation only in mutually-beneficial relationship to itself, as a symbiosis.

In physics this can be seen in what is called wave-particle duality. Wave-particle duality postulates that all particles exhibit wave *and* particle properties, a central concept in quantum mechanics. This phenomenon has been verified not only for elementary particles, but also for compound particles like atoms and even molecules, though the scientific community is still debating whether particles exhibit both properties simultaneously or must be observed one way or the other.

All around us in this seemingly empty space, particles are appearing and disappearing, opposite forces or charges, like yin and yang, which cannot exist without each other. The negative charge is the feminine energy of Creation while the positive charge is the male energy. Though these energies are not genders, they express as genders in the realm of form, hence the expression "opposites attract."

Referring back to the Yin-Yang symbol on page 48, once again take note of the white circle within the black half of the symbol. Even you contain *both* feminine and masculine hormones.

This dynamic, originating force of Creation, remaining one, expresses as two forces and is present on every level of life and the entire spectrum of Creation. The Yin or feminine primal force is unlimited possibility as potential in formlessness. The Yang or masculine is *consciousness* as thought, will or intention.

The interaction of the inseparable and interactive forces very much resembles the act of making love in humans in that the positive and negative charges collide as consciousness penetrates the womb of Creation as formless potential, creating massive amounts of holographic energy under pressure. This is known as the Big Bang which originates from within a Black Hole.

However, the Big Bang didn't happen millions of years ago. *The Big Bang IS happening.* Every act of Creation is a Big Bang and is expressed as the eternal Alpha and Omega, therefore...

Creation is form and formlessness simultaneously. **This is the Divine Paradox.**

The Divine Paradox, or dichotomy of life, is expressed everywhere you look and will be

53

delved into further in this book. Even all humor is rooted in irony which is an aspect of Paradox. The Paradox is the result of the very nature of Creation as Infinite Possibility which must exist in every moment. This means that all exists in every moment in perpetuity, or eternally — the past, present and future all exist simultaneously NOW.

The realm of form, which is the Relative, cannot exist without the realm of formlessness, known of the Absolute, and vice versa. In fact, there is no separation between the two that exist as one.   This is the greatest and also, "paradoxically," least understood symbiosis of all.

Creation in Perpetuity

The resulting energy of the atomic union is more Creation. Scientifically, what is known as fusion occurs upon the excitation of the electrons, initiating a clockwise spiral, creating an electromagnetic field that takes on the shape of a torus, similar to a doughnut. The spiral rotation, pushing the electrons toward the outer edges, opens up a vortex in the core, until centripetal force pulls the electrons back toward the core from both the top and the bottom simultaneously into a "funnel" that compresses itself down to a black hole.

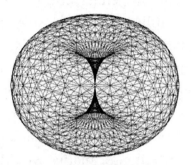

Fig. 5 Torus

Under extreme conditions of heat and pressure, the electrons fuse within the Black Hole. The energy is released as a Big Bang, which is Creation Itself; black holes the formative element in galaxies, rather than their demise. Big Bangs eternally exist on the micro- and macro-levels of the Universe and beyond, which gives cause to temperature fluctuations throughout the Universe, including our own solar system. The greater the energy, the larger the vortex and resultant hologram in its core becomes.

The Big Bang theory of the Universe is misunderstood in that it is viewed by many as a singular event happening in time. If this were the case, the fully-formed galaxies and quasars in the distant "outer edges" of the Universe would be many millions of years older than the 13.8 billion years ago the Big Bang has been estimated to have occurred. There would be evidence of residual gravitational micro-waves and the distribution of matter would be evenly scattered around a central explosion.

Though *like* a Big Bang, the formation of the Universe appears as an "explosion" of consciousness in its entirety as holographic form—*time being created with and within the Universe.* In other words, time is an aspect of the Universe within the 3D structure that is experienced at the level of consciousness only in 3D, though the existence of the Universe itself is outside time. Time is not experienced in the 1st dimension, nor in the 2nd dimension or even the

5th dimension. Neither is space. Space and time are *experiences derived from witnessing relativity* in the third dimension.

The Universe and all of Creation are outside time, including you, as the eternal Alpha and Omega. In reality, nothing is ever created or destroyed. All of Creation's infinite expression is present in every moment. You call it "possibility" rather than expression, as you are unable to witness all the parts at once, essentially rendering what you can't see a "mystery." Birth as the every-thing, death back to no-thing, all multi-dimensional consciousness and all of Creation occurred/occurs/is occurring in the same instant rendering life and death, the Absolute and the Relative, ONE and the same.

Dimensions themselves are nothing more than "levels" of consciousness to be experienced. The fourth and fifth dimensions would consist of your consciousness expanding further in that you actually do experience the past/present/future simultaneously, leaving the linear time plane, though also able to cognate it.

Chapter 5: The Divine Paradox

The Big Bang itself, along with time and space do not exist as a reality but as an apparent condition relative to two points in Creation from a limited perspective. This can be cited in other examples as we call part of the day night or cloudy though the sun has never stopped shining. In this same way, we can say that even duality does not exist, but is an apparent condition of relativity.

The torroidal spiraling process is evident everywhere in nature. Every life form exists within a spiral torroidal field. Electrons dancing around the individual cell exist within its own torroidal field. Even you are a torroidal field, your chakras and aura part of the flow within the torroidal field.

This is the true Circle of Life as energy moves through a three-dimensional torroidal field. You can see the spin in satellite photos of hurricanes from space. Even the Periodic Table demonstrates a spiral as the elements increase in complexity.

The significance of angular spiral spin of every torroidal field relational to one another has been largely ignored or insufficiently recognized until very recently. That the earth is tilted on its axis rotating in an elliptical orbit around the sun, our solar system, within an elliptical rotation around our Milky Way Galaxy, demonstrates the multiple levels of nested recursive processes; all looping energy, what we think of as cycles. Space and time is part of the torroidal process, from the galactic to the subatomic level of Creation. And this energy is self-maintaining and self-sustainable due to the spiral.

The nested torroidal flow patterns, self-similar and self-referencing, set up a stable vibrational resonance from their mutually-beneficial relationships to each other that generate charge and spin. The charge and spin is what creates the atom in wave form. Again, this all appears to be happening in time, but time is the experience by Creation as you, self-recognizing the self-organization of yourself. In 3D, Creation's own fractal organization is witnessed linearly to create the apparent condition of time.

Chapter 5: The Divine Paradox

To call form and formlessness opposites would be to place a divider of separation smack in the middle of Creation when there is no dividing line and there is no center as all is interconnected. This is also what the Yin-Yang symbol is attempting to metaphorically convey within a 2D figure.

Form exists within formlessness, within form, within formlessness. Nested torroidal flow patterns beget nested torroidal patterns as man and woman beget child. From the complex down to a simplistic, metaphoric apple with a core and its "seed of life." Our Universe within the third dimension and its conscious life forms are pretty simplistic in nature. Can you imagine that more complex life exists within the totality of Creation?

Take a deep breath in and then out. The Circle of Life is the inhale and exhale of the breath of God. See how the breath returns to Source like a river returns to the ocean, the tipping point from mortal to immortal inevitable. Not the primacy of your existence, all thoughts, experiences, even your life is a temporal experience returning you to your very Source. And what is the purpose? *Every-thing exists so that no-thing can exist.* The ONE can only BE and know I AM through the sum of its "parts".

This is the Divine Paradox.

CHAPTER 6

The Case for God

Scientific inquiry is impersonal, necessitating objectivity and causal determinism, representing left-brain activity. Equally valuable is the abstract that arises out of right-brain activity through the expression of religion, spirituality and all other metaphysical exploration rooted in the quest to discover the very purpose of life.

However, Quantum Mechanics has knocked loose the very idea that objectivity even exists as consciousness is seen to play a role at the subatomic level as scientists seek to unify the two great theories of Physics.

While Max Planck is the German theoretical physicist considered to be the father of Quantum Theory, Albert Einstein and several others also established foundational Quantum Mechanics, known as the theory of the very small, in the first half of the 20th century.

Chapter 6: The Case for God

Quantum Theory deals with physical phenomena at the microscopic level of atoms and subatomic length scales. This is where we observe paradoxical wave- and particle-like behavior and interactivity of energy and matter and where mathematical formulations become abstract.

As we reverse direction once again, moving from the theory of the very small to the theory of the very large, Einstein's Theory of Relativity pushes through the limitations of Newton's laws of motion by positing that all motion is relative to spacetime, space and time being considered relational to one another — the fourth dimension.

The Theory of everything or final theory that scientists seek would unify both theories and fully explain and link together all physical phenomena. Except that a very large "elephant in the room" remains. Scientists have yet to fully understand what creates mass from elementary particles.

Yet, all roads lead us to a singular desire held by every human being on this planet, represented by the search for the "God particle."

Is the Higgs boson particle the interface between the material and non-material worlds?

What's fascinating is Einstein's formula $E = mc^2$. E is the energy of (m)ass x c^2, the speed of light squared. Light is actually information, "Let there be light." Information is transmitted via light as (c)onsciousness — though Einstein actually meant c to denote a (c)onstant value of the speed of light.

Space, literally, is the final frontier, but not in the context originally held when those words were uttered. Empty space is not actually empty. The spark of Creation is Creation Itself; an energy form as *consciousness*.

As the phenomenon of bacteria and virus, living in the realm of the unseen, once was a mysterious force to mankind, humans now embark upon the final frontier of Consciousness as the all pervasive unseen "elephant in the room." All matter exists in an ocean of consciousness, the "blue touch paper" that precipitates it.

Chapter 6: The Case for God

The conflict between science and religion originated from whether or not the laws of Creation are fixed. Science says if the laws of Creation are fixed, then an outside Creator is not necessary. Religion says if the laws of Creation are fixed, it does not allow for the mystery and miracles of a loving God that offers humanity redemption.

What if both sides of the same "brain" are operating from an equivocal "original sin?"

When Dr. Hawkins stated, *"...it is not necessary to invoke God...,"* was he operating from the presumption of a *separate God*? Has religion so thoroughly removed humanity from its Source (except for a miraculous few) through its primitive anthropomorphization of God that we've forgotten foundational Truth?

The Absolute and the Relative co-exist as the ONENESS of Creation via the unifying field of *Consciousness.*

The Creator and its Creations are one and the same. There is no line of separation. No beginning, no end. The illusion of separation is the veil of which we speak.

When an atom in wave form is observed by consciousness, a coupling takes place, the center womb of Creation, or Vesica Pisces, begetting the atom as particle. Creation begets more Creation just like human yin and yang beget child. Recall the two circles and the Fibonacci sequence:

0 1 1 2 3...

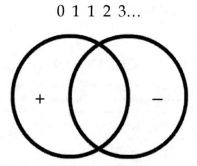

Fig. 3 Two Dimensional Figure

0+1=1, 1+1=2, 1+2=3...

The number series could be also referred to as *dimensions* as we continue the series: 2 + 3 = 5th dimension. (The 4th dimension is spacetime.) Notice how the two dimensional figure above resembles cellular division. You can also view how the overlapping 2D circles resemble a sliced, cross-section of the 3D torus, the central Vesica Pisces giving birth to form in holographic form.

Let's refer to the Chart of Creation for the big picture. It is the *coupling* between the dimensions that begets the next dimension. The nesting of all dimensions within each other is symbolized by the lotus flower.

CHART OF CREATION (Rev. 05/13)

Dimension 0	Dimension 1	Dimension 2	Dimension 3
Absolute	Relative	Relative	Relative
Absolute	Yin-Yang	Duality	Duality
Absolute	Masculine-Feminine	Vesica Pisces	Torus
Absolute	Positive-Negative	+ / -	+ / -
Emptiness	Consciousness	Wave	Particle
Formless	Thought	Word	Form (Deed)

Dimensions are energy forms as levels of Consciousness. All of Creation is a result of sexual exchange as Creation ceaselessly couples with Itself inter-dimensionally. See if you can now name the well-known pattern in the following numbers sequence.

Out of the Absolute originates dynamic first dimensional masculine/feminine energy coupling with Itself, forming what is known as the Vesica Pisces or Womb of Creation—the second dimension.

The second dimension may be where the Higgs boson resides, as scalar field-like potential energy. Lacking spin and charge, the second dimension couples with the positive and negative charges found in the first dimension.

Begotten is third dimensional form, borne under intense heat and pressure. Can you see the pattern of 0, 1, 1, 2, 3...PHI or the Fibonacci series?

Call to mind the looping spiral torroidal flow pattern as one that first pushes the electrons to the outer edges, before centripetal force pulls the electrons back towards the core, and we are able to see how polarity arises. The outward pushing motion would seem to describe what we call dark energy, which is pulling the

Universe apart. As expansion hits the curvature of the torroidal field, polarity begins to disappear as the electrons turn inward toward unification or zero point.

As the negative force of the electrons collides, creating massive amounts of energy, mass appears as the positive counterbalance, generated out of no-thing. On a macro level, this demonstrates how the Universe can be generated out of nothing in the same way that protons seem to appear out of nothing on the sub-atomic level.

Centripetal force, or gravity, created by the torroidal flow is the precursor of mass, rather than the result of it. If gravity arises from revolving energy, then perhaps our understanding of Newton's second law of motion is inside out?

The expansion/contraction paradox once again allows for both axioms of "opposites attract" and "like attracts like" to exist harmoniously in our physical world. Love/hate, light/dark, good/bad, hot/cold all exist in that one could not experience love without hate, light without the dark or good without bad.   Even the human body is an *experience* of being "other" within the totality of Creation.

Opposites are a perception derived primarily from the left hemisphere of your brain that allows the "you" to declare "I am."   Your limited perception, garnered through your senses, strengthens your belief that you are a separate being from the energy system surrounding you.

Experiences are based on relativity between two pretty narrowly defined points.   For example, the outdoor temperature can drop twenty degrees in a day and you may say the weather has gone from warm to cold. A coffee table may be sitting in front of you; it appearing to be "separate" from you due to the space in between you and it, yet, if you pull back and view the entire scene, you and the coffee table are particles of the same "movie" set.

The fourth dimension is where we begin to see duality merge again and how dimensions are nested within each other. This very much can be seen through the 2D Möbius Strip, named after German mathematician and theoretical astronomer, August Ferdinand Möbius, who made the discovery in 1827.

Notice how a **one**-dimensional strip given a half twist begets the **two**-dimensional object in Fig. 6. You can also view the spiral or fractal nature of Creation within this shape, which exists even in the very strands of your DNA.

Fig. 6   Fig. 7

Mobius Strip   Torus Shape

Now observe how the two-dimensional Möbius also *nests within* the **three**-dimensional torus in Fig. 7 on the previous page, with its same topographical curvature. We now have space and time. See how the Mobius Strip becomes the infinity symbol in Fig. 8 on the next page.

Fig. 8

The Consciousness of Creation, what indigenous peoples call Ether, is the universal medium connecting and infusing the totality of existence as the defining force behind all matter.

All that transpires in the seen world is symbolic of Consciousness. This is where everything observed and processed through consciousness, whether on an individual or collective level becomes relevant. This is where all "meaning" and "purpose" is derived— through subjective experience on a local level, though all of life from a state of non-local totality has no other meaning than to exist; to BE.

All in the seen world is a metaphoric mirror *image* of ONENESS of Creation— "...and God created man in his own image" Gen 1:27—

through its infinite number individuated facets, as Consciousness made manifest—thought. The realm of form in the Relative is metaphoric as Creation is formless in the realm of the Absolute. The no-thing of the Absolute remains no-thing in that the realm of form is all a unified *hologram*. If it were not, separation would be real. We know this to be true as the collision of positive and negative particles with the same mass always equals *zero*.

Take a look in the mirror so that you may perhaps grasp the truth in this. Is the image staring back at you, who is able to laugh, cry, show emotions and even age, you? You are physically in form, and appearing as a mirror image, all the while only one of you, not two, exists. The mirror itself is still the mirror, remaining unchanged, while all three—you, the mirror image and the mirror—all exist simultaneously together. Again, pull back from the scene, toward a larger point of view and see that all aspects exist *in and as* ONE.

This example is in itself metaphoric, yet may help you to grasp all that is being presented here.

Creation Begets Creation

Even in *zero* Creation exists as a formative and self-generating force. When the negative electron collides with a positron, both particles are annihilated, yet produce gamma ray photons. Gamma ray photons produce *consciousness*.

Research submitted by UC Irvine astrophysicists to the American Physical Society journal, *Physical Review D,* detailed that between August, 2008 and June, 2012, NASA's Fermi Gamma-ray Space Telescope orbiting Earth found more gamma-ray photons coming from the Milky Way galactic center than they had expected.

Let's examine your dreams. Your physical body lies in a state of sleep while your consciousness travels anywhere in the world or even flies beyond. You experience intense emotions, give birth, lose loved ones, die, you can even orgasm. You wake up breathing hard, sweating, in a state of terror or perhaps euphoria; the experience just as real as if you had lived it outside the dream world. Your body felt just as solid and your emotions were just as authentic as when you were awake.

Yet, as the dreamer, you were always completely safe and it was never possible for you to actually die. And that body that felt so real was part of the experience of the dream. Ask yourself, are your dreams separate from you or contained within you?

Creation is the dreamer. And what of the dream? Every image and experience you have is designed to be a giant arrow pointing consciousness to look beyond the mirror image to come to know Thyself as the ONE, the Eternal Absolute as formless Awareness. The fractal nature of Creation is like the many facets of a diamond and acts AS mirrors unto Itself. If you are *part* of the ONE, then:

There is no outside intelligence judging its creations, granting some miracles, withholding others like Santa Claus. *God is Creation Itself and appears as Life indivisible;* ALL Life, as every thing in the world of form and *also* formlessness, as no-thing.

The Absolute cannot exist without the Relative and the Relative cannot exist without the Absolute. Nothing is born that was not first conceived as an idea or concept derived from a thought that is Consciousness, also known as Qualia.

Moving further on the fractal scale of complexity, the Universe's overall hologram, in its entirety, is maintained by the Consciousness of the Universe. Creation Itself is a giant toroid, its very center the still point of the Tao; the primordial non-dual ground of Being. This is where we are being pointed. This is where union with God is found. It is found nowhere else than within.

Thus, your body, your life, your emotions, your "story" is all an *experience* in holographic form rather than actually who you are, precipitated by Creation Itself. Who you are is the nameless ONE even beyond thoughts or parts as the subjective experience of Creation, just like the existence of the Universe is outside the existence of space and time.

It may be very difficult to digest that Creation/God is the light and the dark, or what is thought of as evil, until you understand that what appears as light and dark on a narrowly defined local level turns out to be something entirely different on a non-local level or from an expanded point of view. Recall the parable of the elephant and the blind men and that duality is a condition of relativity rather than a condition unto itself.

As a small child you may remember being told that God is everywhere. Certainly Creation/God is everywhere as the seen and the unseen. Somewhere along the line, a disconnect occurred in that you were then taught a very different story of a separate and vengeful God. Truly, primitive man projecting anthropomorphic qualities of the Relative onto the Absolute. In simple terms, man creating God in man's image.

In fact, the stories you've been told have so many holes in them, it's a wonder a thinking person could swallow them without the promises of extreme penalty and/or reward offered by blind obedience.

Yet, this is not about assigning blame. Perfection lies in all things. Humanity has passed through many "hallways" of ignorance in its evolution toward the fullness of truth and the recognition of who we are, what God is and our role in the Universe.

At one time, in our primeval understanding, we even sacrificed animals at the altar, believing that to be what an outside God needed to be pleased with us. How many wars and horrendous crimes have been committed in the name of God and continue today in our misunderstanding of what God is?

The dark, in the service of the light, is Creation forever pointing you to experience a larger, more inclusive and holistic version of yourself. More complex fractals of Creation allow larger amounts of Creation to know Itself via the metaphorical information contained in the light or photo particles It projects. Thus, "let there be light" is truly Creation shedding light

79

upon Itself. Creation is Self-referencing, Self-recognizing and Self-organizing.

It is when Creation is witnessed from a single point of light within the 3D Universal structure that the co-existing harmony of the remaining parts gives the appearance of space and time via the coordinated processing of the left and right hemispheres of the brain. Spacetime is created within the brain as it organizes limited incoming information as a precipitate of relativity. We now have the appearance of the conditions of evolution and even cause/effect, though all of Creation exists in and as the eternal NOW. Evolution, the structural Self-organization of Creation rooted in relativity, though actually causal, and Creation are one and the same. We can stop arguing about it now.

If evolution is really just the experiential fractal scale of Creation, are humans guilty or innocent? What of original sin? If there is no outside God, who or what will save us? How can God be evil? How do we get to Heaven?

CHAPTER 7

Intermission

Let's pause on our sojourn for a moment and just breathe. Right now, it may feel like the rug of your world has been ripped out from under you to learn that no one is looking down and watching you.

In a seemingly dualistic Universe, it is by experiencing what you are *not*, through relativity, that you become aware of what you are. There is no better catalyst than suffering to propel you and the rest of humanity to expand our gaze.

Once you understand what is happening, it should take some of the fear away, which is the main component of suffering. It is the unknown that we fear most and by shining the light into the dark, we see the dark is not an entity unto itself, it is powerless.

Ironically, it is the light that is feared more greatly than the dark. For many, personal identity found in one's own dark misery contains a known familiarity versus moving away from addictive behavior, into the unknown of the light—who am I without this looping thought pattern called ME? This is why so many cling to traditions, beliefs and ideology that has long ago/since quit serving their highest and best interests and are now actually causing harm and suffering.

It has been the overdevelopment of the left half of the brain, or yang, in our culture in recent history that has led humankind to expand disharmoniously with planet Earth. In turn, there is an imbalance with the yin or right side of the brain which connects us with the metaphysical or non-local aspects of Creation. The Pineal Gland in the brain is the interface between the two hemispheres.

The right cortex of the brain processes your non-linear, abstract perceptive abilities along with creative impulses, facilitating enjoyment of and participation in pleasurable pursuits such as music, art, and literature—all extensions of Creation. This all-important cortex allows you to tap into and even transcend to the non-physical aspects of your being.

The loss of these "electives" in our education system has far greater ramifications than we realize when we forgo an entire aspect of our being in the quest of short-term productivity. We can look back through recent human history, essentially a nano-second of all time, and recall other periods of human-generated disharmony that birthed renaissances and a renewed commitment to end the exploitation of each other.

Even if humans stubbornly refuse to consciously and voluntarily make necessary adjustments toward sustainable life, Life precipitates conditions unbearable enough that a collective human cry is heard as the symbiosis of Creation continues unabated.

If this sounds like Creation/God is cruel, then first know that as you are infinite in nature, it is all just a dream and no one is ever harmed. You are being called to awaken from the dream of separation and to return to a state of balance/harmony so that you may remember the expansiveness of your true nature.

Also, know that as soon as you come to the fullness of awareness that suffering is no longer useful in the experience, then it falls away, regardless of the circumstances you find yourself in. This is generally not accepted at face value. It must be experienced.

It is through the sum of the parts that Creation/God is aware of Itself in its entirety. Consciousness exists in the realm of form as a tool of recognition of what is beyond consciousness — as Awareness. It is *through* consciousness and its denser quality, form, that Creation can fully be aware of Itself as an *experience*.

This is the value of experience, to generate the expansion of consciousness (in the body) which in turn generates awareness (beyond the body). This is what is meant by awakening from the dream, to become fully conscious and aware. To become Awareness Itself is to enter a place of power rather than force; a life of love, joy, peace, freedom and abundance rather than one of human-created hardship.

Love is the substance of all Creation/God and you are part of this endless ocean of exquisite beauty and eternal bliss. How do we know this?

From those whom have transcended the human experience into awareness of the truth of who they are in totality — the Masters and Sages. We even experience this bliss on a daily basis when we slow down enough to lose ourselves in the natural beauty that surrounds us. We experience ecstasy when we lose ourselves in the arms of a lover, the lover really reflecting back your own nature as love itself.

It is understandable that there being no outside deity is difficult if not impossible to fathom in this moment. For many, your entire way of being is based on a power higher than yourself — a symbol of hope for a brighter tomorrow — as an escape from what appears to be a harsh and unforgiving world filled with hardship.

85

Chapter 7: Intermission

You are safe. Nothing is going to happen to you because you have read these words, because you've questioned God or your local authority. You have not blasphemed Creation/God in any way. Personal inquiry with the inclusion of scientific reference does not automatically make you an atheist or agnostic. Rather, inquiry arises naturally when you are ready to go deeper into the depths of your being.

Science is one of the many paths of inquiry as are the arts, religion, spirituality, walking in the woods, emotions such as fear, hate, loneliness, hope, faith—from the macro to the micro—all tools that are useful until they are not. Then you will notice the need for them disappears from your life. In reality, there is no activity/symbol in your experience that is not pointing you toward the singular Truth of who you are as the Absolute.

Self-experience is the source of Creation's very existence as Creation can only know Itself in terms of "other." As we mature and begin to realize our holistic nature, inquiry leads to shedding the stories told to children by those also spiritual children, or perhaps misguided individuals. This book's purpose, as a "blue touch paper," is to ignite your own intuition through inspiration, leading you toward the fullness of who you are as Truth.

What is being offered to you is a tremendous gift to the end suffering by stepping out of the illusion of hope into a higher power — *your* higher power through the richness of Life Itself — and into an enlightened and awakened reality which cannot be found in the future. It can only be found in this moment of NOW.

This is the only true freedom there is.

This is where it gets GOOD....

PART II

CHAPTER 8

The Two Trees in the Garden

"And the LORD God planted a garden eastward in Eden; and there he put the man whom he had formed. And out of the ground made the LORD God to grow every tree that is pleasant to the sight, and good for food; *the tree of life also in the midst of the garden, and the tree of knowledge of good and evil.*" – Genesis 2:8-9 [emphasis added]

In the middle of the Garden of Eden stand two trees. One tree is the Tree of Life, the other is the Tree of Knowledge. The first thing to notice is that in the middle of the *one* Garden stand *two trees*. Let that sink in for a moment...

Both trees are of Creation/God and both trees are part of the same Garden. One tree is a metaphor for Life (the Absolute), the other for Death (the Relative). Remember now that *both* trees stand in the *same* Garden.

Chapter 8: The Two Trees in the Garden

Have you figured out that the Garden is Creation Itself? In the Garden is both formlessness and form. There is no separate Garden. You may need to keep reminding yourself that there is no separation. This should lead you to the recognition that there is no separate Heaven or hell.

Even Pope John Paul II, a great man of enlightenment, stated in 1999 in front of a general audience that in speaking of hell as a place, the Bible uses "a symbolic language," which "must be correctly interpreted ... Rather than a place, hell indicates the state of those who freely and definitively separate themselves from God, the source of all life and joy." Hell is not a place, but an experiential *state* of thought or being.

You are standing in the center of the Garden of Eden, and *both* experiences of Heaven or hell are available to you. Right now.

"In the beginning was the Word, and the Word was with God, and the Word was God." – John 1:1

The Word manifests directly as the realm of form via thoughts/consciousness. Creation/ God is *equally* the formless and form, eternally existing as states of primacy. What determines whether you experience Heaven or hell, eternal life or death?

It is determined by which tree you eat of metaphorically or where you are "looking;" locally or non-locally. You've heard of the expression "food for thought." The Tree of Knowledge represents thought. Thought appears to divide the indivisible Absolute, giving the appearance of the relative or local. Thought is separation itself—in idea or concept *only*. Thought begets more thought just as food feeds the body's cells, allowing them to self-perpetuate.

When thought is personally identified with, you "experience" a "separation" referred to as death. Identification with an original thought begets the next devolved thought; i.e., I am the separate thinker of these thoughts, creating the experience of separation. Metaphorically, man and woman are then kicked out of the Garden where we must "clothe ourselves;" meaning we don form. Of course, you have never literally been kicked out of the Garden as Creation/God is both formlessness and form.

All thought exists non-locally as unmanifest potential. Thought in and of itself does not manifest physically without the necessary ingredient of consciousness. Matter then coalesces as part of resonance patterns, gaining stability, when in the presence of mutually-beneficial, self-similar resonance patterns — the power of two or more gathered in prayer — consciousness. The bottom line is that form only exists if it is beneficial to the whole.

Clumps of resonance patterns give space the appearance of matter being randomly scattered around the Universe as Einstein's General Theory of Relativity predicts.

This was demonstrated back in the '70's via a computer simulation called The Game of Life developed by mathematician John Conway. In the simulation single squares light up in an empty field of unlit squares, and then grow dark again (death). At fast speeds, the single squares keep lighting and dimming until something happens when a single square has three other lit squares adjacent or *touching* it.

Instantly, two of the squares remain lit as the process continues, random squares lighting up faster and faster until three squares are lit within that same clump, and so forth. Meanwhile, clumps are appearing, seemingly randomly in other quadrants of the dark field, which represents the Universe. Begin to imagine clumps continuing to grow in number and size. These clumps represent matter.

What appears to have been random or chaotic was not really random or chaotic at all. This is Creation's Self-sustainable nature, eternally in the highest and best interest of all of Creation—ensuring its own perpetuity. Rather than survival of the fittest, it was the un-survival of the unsustainable—those lit squares that did not match with resonant lit squares. What does this say about human competition?

Competition is self-limiting and ensures its own demise in the same way that if one animal becomes too dominant in the "neighborhood," it depletes its own food supply; its survival no longer sustainable.

Your Role in Co-Creation

It is not until observed by Consciousness that wave particles collapse and the hologram of our existence appears in our local awareness. What are you and what is doing the observing?

You are intimately connected with the existence of your reality as co-creator. This does not occur in the context that a separate individual picks and chooses what you observe or even what thoughts you think. Separation is merely perceived, it is not a reality. Who is the observer behind the thoughts? What is observing the thoughts?

Remember, it is only through personal identification with the content that you observe that thought devolves into the belief that you are separate from Life; a separate doer. You are ONENESS Itself. As soon as you believe yourself the separate doer, life becomes entangled and hardship ensues.

Thought, or the Absolute divided, then becomes further "corrupted" when viewed in part rather than from a holistic perspective. The individuated part has not separated from the whole and only appears as such as our singular attention or conscious awareness has narrowed from the ocean to a wave in the ocean to a drop of water. We've forgotten the drop of water is the ocean.

Let's now re-contextualize the meaning of corrupt toward a more holistic point of view. As the nature of Creation/God is the Absolute/Relative, meaning Infinite Possibility expressing Itself relative to Itself, corruption is simply a fractal of the whole of the expression of Creation/God as it continues to divide or devolve.

Even as Creation/God appears to divide again and again, its fundamental wholeness and perfection remains intact, unceasingly. It is important to note that Creation is not actually dividing; only the focus of the point of view is getting smaller and smaller.

Chapter 8: The Two Trees in the Garden

One's focus can become so narrow that perfection begins to be seen as imperfection. Suddenly, it seems as if Nature must be making mistakes.

As humans dwell on their own singular point of view, now steeped in personal identification, corruption further infiltrates the totality of Life as it divides and deviates, divides and deviates. Man and woman come to view each other as separate from one another and separate from Nature. The devolvement from holistic Truth continues until entire groups and sub-groups of humans are labeled a mistake and even marked for extermination by other humans... That is hell.

CHAPTER 9

The Root of All Evil

Belief in separation has allowed mankind to create imagined opposition such as war, evil, judgment... We create it all.

Evil does not come from an outside intelligence or power source anymore than God does. Evil is nothing more than Creation so divided in perception, so distorted from its original wholeness that new meaning has been assigned to the fractal through personification; distorted contextually, whereby you are experiencing such a small piece, or fractal, of the original image.

It is the personal identification with thought that is the veil that creates the illusion of separation. Hell is a state of mind.

What is cooperation becomes competitiveness. What is joyful becomes sorrowful. What is abundant becomes lacking. What is peaceful becomes warring. Love-based behavior becomes fear-based behavior. Can you think of any fear-based behavior on this planet? Do you see that *nearly every* human-created system we have installed is fear-based, including religion?

Only a relative handful of people on the planet have ever been able to free themselves of the mind and its resulting social constructs that have enslaved the rest of humanity. Martin Luther King's words speak of much more than social justice, "Free at last, free at last, thank God Almighty, I'm free at last."

<u>What is Sin?</u>

Thought is the root of all "sin."

The original, most widely used definition for the word sin is "to miss the mark," as in archery.

The Tree of Knowledge is a metaphor for thought. Thought is not Truth, but rather a conceptual and subjective commentary *about* Truth or Creation/God, like a reflective mirror, which can only exist as a holographic projection as "other."

In the same way that a mirror image of you is only fractional, encompassing only two dimensions of your complete nature, the image does not contain all aspects of you such as your brain, organs, chemical reactions, thinking processes, etc. In this way the mirror image is "you," but not the full Truth of you. Even the "you" in 3D does not encompass the entirety of you, as you exist in more than three dimensions as part of totality. As the mirror projection appears temporal because what we observe on the screen of life appears to be changing, it is why 3D is associated with "death." A more in-depth explanation of this construct can be found in the next chapter.

Thought or consciousness creates the appearance of separation on a local level with its power to construct or deconstruct matter in the temporal world of form. Your thoughts neither change nor affect the reality of your true nature as the Absolute, only alter/distort your perception of it as you are only seeing the tail of the elephant.

All pain and suffering are created by mankind's personification of and identification with thought which then manifests as existential guilt. In other words, thoughts continue to beget another wave of thought or devolve away from totality—like ocean waves that appear smaller and smaller farther from their source. As you feel smaller and smaller which is the opposite of who you really are, guilt thoughts lead to self-destructive behavior as we first harm ourselves which leads to lashing out and harming others which is actually a projection of self-harm onto "others."

What is war if not a thought of separation? What is discrimination if not a thought of separation? What is slavery, what is murder, rape, theft, divorce, anger, depression, sorrow, sadness...?

All of the Masters from Jesus, Buddha, to Krishna taught us by example to *transcend* thought, recognizing the natural state of our inherent wholeness as Love; our true nature.

The Buddha meditated under the Banyan tree until devoid of a single thought, reducing the Ten Commandments down to the law of love. Jesus was love's very embodiment and *as* the embodiment of love said he was the "way and the life."

By no means did any Master intend that a requisite to believe in them personally or worship them would have to be met for you to get from where you are to Heaven, whether here on Earth *now* or in an afterlife.

Embodying the very qualities of the fullness of Creation/God as Love, it was inevitable that a suffering woman or man would, like a child, fall at the feet of the appearance of a human being Oneness Itself. Just as a child matures and becomes a sovereign and self-sustaining adult, so do we mature spiritually to discover that all we ever need does not come from outside of us, but from within.

It was later on after Saint Peter and the Apostles set out to spread the teachings of Jesus that the formal institution called "Church" was officiated along with those who wished to manipulate Jesus' teachings for the purposes of power and control. Behind every Queen and King has been a Bishop and behind every Pharaoh the High Priests.

One such method of manipulation is the severe restriction of information, thus changing its context. Another is to provide false, alternative versions. There are three stories that may immediately pop into your mind depending on your culture and heritage.

The first is the history of Earth and of Mankind. There appears to be a concerted effort to hide the actual historical record of humanity's narration in relation to our planet. Architectural evidence, historical documents and more suggest periods of time on the planet when civilization seemed to be more advanced and more connected with Nature than we are now and even the possibility of non-native intervention.

What happened to send us backward in our evolution to where barely more than a hundred years ago comes the invention of electricity and women get the right to vote? There are a lot of humans alive today that are over 100 years old. We are talking one person's lifetime; the use of horses or mules to pull a plow versus the complex agricultural systems that were once part of the fertile Tigris-Euphrates river valley in the Middle East.

And what of the historical documents previously mentioned? Though an unfathomable number of significant documents were purposely destroyed in the fire at the Library of Alexandria, Alexandria was hardly the only document repository in the world. What institutions are holding these documents and why aren't they available to the public?

The idea that all of humanity in all of our diversity came from Adam and Eve who were created by God one day a few thousand years ago is absurd in the least and ridiculous at most—when artifacts and human remains of record indicate a different reality.

The Divine Paradox does provide truth in that the totality of the Universe exists in the same instant.

Chapter 9: The Root of All Evil

Once it is realized that Evolution is Creation in process—*within* the 3D construct—one does not fear the concept that even man and woman are subject to the laws of Evolution. Modern man evolved from primitive man and will continue to evolve to a fully self-actualized version of him- and herself utilizing those organs in the body and the strands of DNA that seem to be unnecessary or "junk."

Neither has man evolved from apes. Creationists nor Evolutionists have it just right. Man is a "special" creation as its own fractal, yet man and woman are not special in and as far as Creation/God is concerned. And no gender is more "special" than the other.

Consider that false and alternative information has been included within the allegory of Adam and Eve. The supposition is that Eve was tempted by the devil himself disguised as a serpent into eating an apple from the Tree of Knowledge. Further, she then beguiled Adam into eating the apple, too, placing humanity in a state of "original sin" and subject to the penalty of "death."

We've already covered that the Tree of Knowledge is a metaphor for the idea of separation that occurs as a result of thought itself when personally identified with. Thought is the Absolute divided or more accurately, viewed in fractal, relative parts, creating the illusion of separation and a change in context from meaninglessness to meaning-full. Each projected thought by Creation/God is but a drop in the ocean of all Creation/God.

Also, both trees exist within the same Garden — the totality of Creation/God and that as humans, it is possible to experience both union (Heaven) and separation (hell). All experiences and form in the third dimension contains a positive *and* a negative charge, making you life's sinner *and* saint.

What if the story of humanity is complete as it stands in the above two paragraphs? Where is the blame, then, for humans being...human? Where is the separation from Creation/God? Where is the judgment, the condemnation, the promise of Heaven if you are obedient, or hell without Jesus or any other Master or guru to lead you there?

Blame, judgment, separation, condemnation, reward, punishment and even death all fall apart as human points of view on a relative and local level that do not encompass the vastness of Creation/God in Absolute Totality.

The meaning that has been assigned to the Adam and Eve allegory has been derived by a limited, human point of view. We tend to take objective reality and lend it our own personal, subjective meaning — often times this meaning told to us by others — passed down generationally via oral storytelling until it lands in the hands of a select, and at times even secret, group of individuals who canonize the official version of the story that if not adhered to is punishable by death.

By re-contextualizing the story of Adam and Eve as a fall from Grace resulting in disconnection from Creation/God, labeling it a wrongdoing, you open up the possibility of a Godhead, a power-over structure, that can be emulated on the human level as dictatorship.

Christine Horner

Further, if the story is tweaked a bit by a patriarchal society, "authority" can eliminate the possibility of half the population from participating in this dictatorship of the mind, by calling out women as the weaker sex, thus eliminating women from any positions of power within political/religious structures and even within their own family unit by requiring obedience and subservience to a superior man.

The Story of Creation

The Story of Creation is a completely blameless one. Please refer back to the Chart of Creation on page 68. In our Universe, or the Relative, Eve is seen as the feminine energy or negative charge that is the formless. The masculine, positive energy of consciousness, in form, is Adam. These energy forms are expressed as genders arising as *part* of the third dimension in the same way that space and time do. Beyond 3D, these complimentary energies are genderless.

Out of the perfection of Unity–the Garden of Eden–in the same way atoms appear out of the

Chapter 9: The Root of All Evil

Void, the idea of "other" issued forth. The coupling of the feminine energy (0 Dimension) and masculine energy (1D) then beget the Vesica Pisces (2D) out of which all form arose (3D). What is separation from Creation/God that we humans are to be blamed and held accountable for?

Separation is space; literally and figuratively. Separation is not an actual condition, but a description of an apparent condition while viewing a limited point of view. For example, electrons are a "half mile" away from the nucleus of the atom, yet the electrons are still part of the atom, the atom remaining fully intact as the *whole*, as the electrons remain as a distinct *part* of the atom. Atom, atom, adam… ADAM.

Are you smiling yet? Man and woman are innocent. Creation is "creation-ing." Life is Life-ing. The true Holy Trinity once again illumines the Divine Father/Masculine, Divine Mother/Feminine and *Humanity*. Humans, in the fullness of our expression as Human Beings—mind, body and spirit—are revealed to be the bridge between Heaven and Earth, as Jesus was called to be.

Jesus said, "Even the least among you can do all that I have done, and greater things." Jesus said he is our *brother*.

Christ on the Cross is symbolic, the cross itself representing the first dimension intersecting with the second dimension, Jesus being third dimensional as Divine form. The Holy Trinity is the aforementioned dimensions; the third dimension just as holy as the second, the first and the realm of formlessness or the Absolute!

Once you relieve your mind of the limitations placed on man and woman by human thought, there is nothing left but forgiveness and even that isn't necessary once you allay yourself of the need for forgiveness. All illusion falls away to reveal that Creation/God is...Love.

Love is what Jesus, Krishna, Buddha, King, Lennon, Yogananda, anyone you can think of, espoused and taught us to BE, because they were. The great I AM is a form of the verb "to be." When Moses realized Creation/God in nothing more than a burning bush, the light of the fire representing illumination that comes from awareness, "I AM THAT I AM," his conscious awareness expanded, in recognition that Creation/God exists not only as the formless, but in all form. This is Creation/God recognizing Itself through the Self-actualization of man!

For millennia, humanity has struggled to understand one simple truth. There IS ONLY God.

Again, this is the Divine Paradox. Creation/God is the Absolute AND the Relative, the whole and the part, the non-local AND the local, the light and the dark. There is no place and no-thing that Creation/God is not. It is impossible for the human mind to comprehend this. This is the Great Mystery spoken of by so many prophets, sages and all illumined beings.

There is no separation, no distance and no time. The illusions of separation, distance and time are nothing more than thoughts. All form is but empty projections of thought by Creation/God; a grand dream that includes you, in the same way that you dream the fullness of an alternative reality while you sleep, the five senses seeming just as real in the dream as they are to you right now. These are thought-forms.

And in the dream, from a holistic point of view, all things are equal. There is no better than, higher than, much less good or evil, though you experience what appears to be a range of emotions...infinite points of view that make up Creation/God recognizing Itself. It is Creation/God looking through your eyes as the observer.

Attachment/separation, life/death, coexist as Oneness in the same holy instant of eternity, though they appear to be miles apart. Both imply time/space as part of the nature of those thoughts, but don't literally exist, in the same way spacetime exists *within* 3D, but not beyond it. Even the concept of eternity is time-bound. A more inclusive Truth would be Creation is, by Self-definition, Infinite. Non-linear thinking is required to recognize the true meaning of this.

113

Chapter 9: The Root of All Evil

It is in *recognition* of union, as the observer, that there is annihilation, removing time and space. When you let go outdated constructs, all falls away to reveal the same Oneness of Love. The eternal now remains unscathed and is sentient, immortal and omniscient, as are you.

From your vantage point, look down at your exquisite planet Earth. How could you ever look at a rainbow and say red is better than yellow or green is better than purple? Or that the thunder and lightning and driving rain that brought it is any less perfect or separate from it?

See how the cerulean blue oceans hug the green masses of land, how the colossal cloud swirls dance over the curvaceous horizon. Fall in love with it all over again. Fall in love with yourself and every "other" person you encounter! Remember how you were born as Pure Love and didn't hold back the love that you are until you were taught to.

CHAPTER 10

The Nature of Creation

You are blameless. You are Creation Itself, lost in what became an individuated story of separation—the veil. All story is formed by Creation on the human level, or 3D, with correlating subjective meaning and purpose through the creative power of thoughts. Creation is not separate from you anymore than a wave is separate from the ocean.

In the Absolute there is no meaning and there is no purpose other than to exist. There is nothing but emptiness, no subject, no object as a basis for relativity necessary to derive meaning and purpose from.

Creation, whether form or formlessness, just IS—because that is the Nature of Creation, to BE, in all Its Infinite expression. Holistically, it is completely impersonal. You only *think* you are having a personal life. Literally. You only *think* that others, or even arrogantly, you, are in charge.

Yet, even down to the local, fractal level, perfection remains. This is the Divine Paradox in action again that is impossible for the human brain to grasp. The Absoluteness of Creation whether witnessed holistically or fractally, remains untouched and unharmed as projections of thought, which is the Light.

As we finish out the story of Adam and Eve, we discover the narrative is incomplete. Eve bites into the apple which will bring her death. But, inside the fruit is an apple seed, metaphoric for the Seed of Life. A vortex motion splits the seed to create the Egg of Life. The cells further divide to form a more complex structure called the Flower of Life. The Flower of Life symbol is found in most every major religion in the world.

Can you see the 3D shape that emerges from the Flower of Life in holographic form? Look for a cube in Fig. 8.

Fig. 8 Flower of Life

The Flower of Life's hexagonal shape contains the patterns of Creation. Thirteen circles emerge from the "Great Void." This is the source of all that exists, called the Fruit of Life. From this pattern, it is possible to create any molecular or living cellular structure that exists in the Universe—essentially, any living creature. Thus, death generates life, or formlessness generates form, in a never-ending circle, form returning to formlessness.

Every-thing is simply creative thought, an energy, which appears as light out of the void, like a Black Hole; a vibration which is also a musical tone. Think of a Creation as a movie

117

projector and thoughts as endless frames of film playing out as light and shadow. Though the drama contains a myriad of events and sensations from despair to ecstasy, no one is ever harmed in "reality."

The Adam and Eve story is the story of how life and death or the Absolute and the Relative are one and the same, in the same Garden. The "pain" associated with childbirth or the birth of humanity is symbolic of the pain and sorrow in the dream of separation made manifest. It is the story of Creation in all its wonder and mysteriousness, so simplified, a child could understand it. Where is the "sin," the blame or the guilt to be found in this story now?

And What of Suffering?

Suffering is Creation's alarm clock. It is a contraction of awareness or point of view so great you are motivated to seek something more than yourself to end the dream or illusion of separation. You are motivated to seek wholeness. All suffering originates from the existential guilt that arises out of thoughts of separation and the resultant behavior of those thoughts.

Nearly every person lives daily with minimally an unexplainable low-level anxiety you may account for as stress you attempt to distract away by keeping busy—TV, food, sex, work, excessive exercise, computers, drugs or alcohol, video games. The quickest way to verify this is to eliminate all video viewing for one week and observe your thoughts and feelings.

Suffering is life's catalyst to motivate you to expand beyond your personal story by inquiring first outside yourself, in the material world, for assistance. We do this individually and collectively. You turn to friends, medication, social services, support groups and family. Collectively, we create charities, think tanks, governments, leagues, civic and peer groups. And when these don't work, we form more.

Humans, to this day, diligently continue to attempt to expand awareness mostly via the outside world of knowledge and/or by seeking intervention from an outside God or savior for the solution to their problems.

Can you see Einstein's definition of insanity somewhere in here? Doing the same thing, expecting a different result? How would you rate the quality of life for most people in the world right now? Is it joy-filled and peaceful? Relaxed and abundant? Are things improving or seemingly falling apart? Is it Heaven or…something in between? The good news is this is changing.

Based on your current level of awareness, you do the things you do and think the thoughts you think until they are no longer useful. As part of the circle of life, all structures rooted in the temporal world of form are designed to pass away, returning you to fullness of the Absolute.

Emotions, ego, fear, anxiety, time, space, etc., are all *tools* designed to motivate you to look again. In other words, not only can you observe the devolvement of Creation as smaller and smaller fractals, you can return to the bliss inherent in unity by looking the "other" way, from the external to the internal, where you discover your nature as pure awareness—no-thing.

So, too, does freedom begin with a thought. First, there is deconstruction. Nowhere is this more evident in the world than in this moment as all of our separative, fear-based social, economic and political structures crumble. Perhaps the deconstruction is happening in your personal life or you may have friends or family who are in the process of losing everything.

Humanity is presently undergoing an accelerated and collective shift in consciousness which is reflected in our outer world. The evolution of consciousness has always existed. We are just "conscious" of it now.

After clinging desperately to existing known methods and systems of problem solving, you finally become aware that a new approach is necessary. What used to work, no longer does. Levels of consciousness such as pride, competition, patriotism, faith, and hope, which at one time served you, now begin to show cracks as you begin to see these constructs are still rooted in separatism—the negative charge

121

of the 3D construct. For those who have not yet reached these states of consciousness, they will cling to and defend these ideals until they, too, no longer find them useful.

Awareness is the agent of change itself.

This is Life Life-ing under its own power. It only appears that a "you" is drawing conclusions, forming opinions and making decisions. Creation is Self-recognizing and Self-organizing. As part of that, you are not separate from life — *you are Life.*

It only appears that there are lessons to be learned and that you are being compelled to expand to a higher consciousness — for the purpose of creating a seamless "mirror." To have the 3D "you" never be compelled to look toward the totality of your nature as Creation would be tantamount to putting a grid over the mirror and dividing it up into quadrants. Creation would then only experience bits and pieces of itself, rather than experiencing both bits and pieces *and* totality of its no-thingness as love. Creation would then be unable to experience itself in its totality, or be Self-aware.

Also, evolution means the "passing away" of the lower consciousness, bringing form full circle back to formlessness. The "death" of a concept is a simultaneous birth of the greater consciousness.

What we call evolution is Creation witnessing itself fractally; expanding and contracting concurrently via consciousness, an energy form which is like the unseen wind that circles the Earth. Have you figured out that consciousness is a tool, like a mirror? Consciousness is the "human" part of "human being" while awareness is the "being" part of "human being" that is beyond all form. Understanding, growth, maturity, knowledge... all fractal levels of consciousness BEing.

On the local level it only appears as if you have free will and that you are independent and separate from all of Creation. But ask yourself where your thoughts are originating from? Who is behind the thinker?

What does this mean for my life? Am I just a puppet or an actor playing a role? Or am I being played?

Chapter 10: The Nature of Creation

Those types of questions are coming from the ego/personified expression of life. There is nothing wrong with this point of view or these questions. They only lead to expanded awareness that is more inclusive in nature. Once personification is seen through and it is seen that you are Life Itself, Creation Itself, you are no longer bothered by these questions.

Holistically, you now understand that all consciousness/thoughts/things originate from no-thing and return to the Absolute. The purpose for the appearance of death in the 3D realm is to reveal that Creation is also no-thingness. If death did not appear to exist, how would you or Creation have any idea of your/Itself as no-thingness?

Even you are a thought of, by and as Creation made manifest as part of Infinite possibility. What is even more mind blowing is that cause and effect do not exist.

If Creation exists as infinite possibility, then wouldn't infinite possibility be expressed in *every moment* by its self-definition? Reframing, would there be any time that everything/nothing would not exist? Creation, in its totality, would always exist *now, eternally.*

As this is difficult to communicate, much less comprehend; let's go back to our example of the film projector. There appears to be movement, a linear timeline and events. Yet, examine the filmstrip. Each frame of the movie represents a part of the one whole movie. Additionally, each frame also stands under its own power as whole and complete.

Once the movie has played in your personal theater of your life, you call it the past. The movie, as a whole and even each of the individual frames, still exist(s) since Creation expressing as infinite possibility in every moment of now is eternal or rather, infinite; eternal being locked into time. As life is a continual movie, the next "installment" is playing/ready to play. You call this the present and the future.

Each frame, movie, or franchise of movies called your life, from birth to death, whether you call it the past, present or future exists as the eternal now. Can you see how life expanding from one frame, to one movie, to one franchise is fractal in nature? And that the only "movement" is awareness?

As Creation could not exist without perfect harmony and the human realm is relational on the local level, of course the frames of the movie are self-organized to give the appearance of linear time. We know that time is not real when we talk to our friends halfway around the world. It may be 12 hours difference, but if they were truly in the future, we could never "catch up" with them to have a chat. We connect with them in the *now*. It is always now.

As to why, then, can't we communicate with those from the past or the future; it is beyond your current fractal of human consciousness. As you expand into and as more inclusive, larger fractals, the appearance of limitation drops and these possibilities are experienced. In the past, these persons who reached higher states of consciousness before the collective mass were burned at the stake. Today he or she gets her or his own reality TV show.

Who You Really Are

You exist multi-dimensionally; something more and more humans are becoming aware of. In 3D, you are very much subject to the laws of time and cause/effect, feeling much frustration at the limitations you bump up against.

The theme of manifesting your reality has become more and more mainstream, introduced into pop culture the last couple of decades, promising you relief from human limitations. The power of positive thinking and visualization bring a few results, but not the rich rewards we are promised in the seminars and books. Why?

In 3D, your culture and conditioning have laid a foundation of limitation through your belief systems as part of the 3D construct. Think of your daily life and see how you are totally hooked into the idea of cause and effect. "I must get up and go to work today so I can pay the rent." "If I don't make that deposit payment, I won't get the apartment." "I need to leave before 4:00 p.m. or I'll be stuck in rush hour traffic." All limitation is rooted in separation expressed as some type of fear—the negative charge inherent in 3D which is beyond your "control."

Every moment of your day is ruled by the rules you give power to. The rules, in and of themselves, contain no power. Beliefs, in and of themselves, contain no power. It is only through "coupling" with your attachment to those thoughts that you give birth to the limitation.

The third and fourth dimensions are experienced as *subjective expressions of Creation*. The subjective expression of Creation is known as the ego. Also a tool, the ego has been called the root of all evil, yet ego serves a very important function in Creation.

The subjective experience of Creation is what experiences time and space, the Relative, on a local level. Time and space continue to expand until the limitations of the subjective experience or ego are seen, where your external expansion or physical expression, of an "idea" is no longer serving you.

Metaphorically, you've hit the farthest point of the spiral curve of the torroidal field that loops you back in toward the center, bringing inward introspection that occurs in the non-physical plane, until you give birth to greater and more inclusive truth. It is this circle of life, infinite birth and death simultaneously at play (thing/no-thing) that is the appearance of evolution, though evolution is not an actual phenomenon in and of itself.

As you individually, and we collectively, move back toward unification in the center of the torroidal field, expansion becomes compression as space, time, energy and matter condense (STEM Compression). Moving up through the dimensions in consciousness, we recognize the fourth dimension not as space and time, but rather spacetime, and a curious thing is noticed.

Time doesn't seem quite so linear, becoming more abstract. Oversimplifying, spacetime is relational, one affecting the other, differentiating itself from 3D space and time. A fourth dimension allows a three dimensional object to be rotated on its *mirror image.*

It is not until your human consciousness expands into 5D that apparent human limitations begin to fall away and that you start to see how your thoughts are manifesting without much time or even instantly.

Yet you didn't "do" anything or "gain" any "special" dispensation—no exceptional knowledge, abilities or powers—to gain entrance into this refined state of being. What transpires is the falling away, or transcendence, of personification, which is the positive/negative charge of Life. As you move closer into union, the center of the torus, you experience this transcendence; literally the annihilation of positive/negative charge toward a more unified alignment with Creation itself.

Essentially, what is attempting to be expressed here is that self-serving desires rooted in separation that would not be beneficial for the good of all, or Creation itself, fall away in higher states of consciousness. This is how Creation remains in perpetuity, why you are always safe and the non-physical aspect of you imperishable.

Personal suffering is brought on by observing the limitations in the external experience you are having and the strong desire for the experience to end. Seeking to relieve the suffering the fractal/egoic state of perception has produced and that the existential experience cannot resolve, we play every last desperate play in the playbook until, we finally are compelled to transcend the human physical experience seeking a more inclusive experience of wholeness. We finally see the constraint in observing life only in terms of the physical, seeking to know the non-physical, non-local aspects within ourselves.

There is not a single man or woman, from the beggar on the street to the scientist, who is not seeking the exact same thing in his or her unique quest—wholeness or (re)union. How does this transition happen?

It happens by and under its own power. Another name for Creation's Self-sustainable nature is *Grace*. First, the concept comes into

131

your consciousness that a more inclusive possibility exists. Then, the strong desire to attain this possibility arises. What each of us truly is longing for is to experience our inherent wholeness. For, to experience it is to BE it. Do you understand? Experience = Creation BE-ing. I AM that I AM. Creation is the all-in-all as Infinite Expression.

Then seeking arises. We become seekers with all of our hearts to be one with all that is, union still being a "separative" thought concept. Union is actually the root of all desire for love, joy, peace, freedom and abundance. This is the core of why we investigate and explore nature, science and spirituality. We long to re-member our true nature as the ONE.

The irony is, seeking is the separation and suffering itself. Spiritual Teacher Eckhart Tolle, author of *The Power of Now* and *A New Earth*, was asked if suffering is required here on Earth or if it is possible to live without it. His response was, "It is required until it is not."

It is out of the sorrow that joy arises; out of suffering that compassion arises, out of desperation that hope arises and out of the darkness comes the light. And even hope only brings us so far. We then have to let go of the limitations found in hope for a more inclusive point of view.

What has actually happened to bring on suffering is that Awareness, your Infinite Self as the formless, has "moved down" the fractal scale of the mirror image of itself, into its parts, that already exists infinitely. At the same time Creation is moving back "up" the fractal scale to observe itself in Its entirety. Thus, Creation is both able to know Itself on the local and non-local level simultaneously.

The metaphor for this expansion and contraction of life is found in our own inhales and exhale which is the in-breath and out-breath of Creation Itself. It must be qualified that this description is entirely metaphorical in nature, meaning it's very limited in scope in that the human part of your being may have a point of reference to work from. Go within yourself to inquire more deeply.

Chapter 10: The Nature of Creation

What we call "evil" is life's catalyst which brings us home; full-circle to the fullness of what Creation truly is in its totality. The entire elephant!

Moving up and down the fractal scale, called expansion and contraction or life and death — are you seeing a pattern emerging here, the pattern of the torus? Expansion and contraction exist simultaneously in you and the Universe at the same eternal holy instant. Nested interdimensional torroidal flow patterns all exist within the ONE, like nested Russian dolls.

What is difficult if not impossible to convey in words, which are also metaphorical, in this Divine Paradox called Life is that there is no individual "you" expanding or contracting, living or dying. It is simply the Whole witnessing and experiencing Itself. Everything/no-thing exists within each other as the ONE. Why?

Creation would not exist without simultaneously existing infinitely as everything *and* as the void from whence all life is issued; no-thing.

Creation is in Its infinite expression in every moment of now because now is all there is. You are never actually going anyplace or anywhere, doing anything, learning or evolving. It only appears as an illusion that you are. Just as your senses seem so alive in a dream, projected by thoughts, your life seems real, but is equally a dream. Who or what are you? You are the dreamer behind the dream as pure Awareness. You are Infinity Itself.

Now aware of what you really are as the Infinite, when contrary experiences do show up in your life, you will no longer regard them as personal and thus suffering is greatly diminished.

Epiphanies and miracles are experienced when the images are catapulted to a larger fractal of awareness as spacetime is bypassed, no longer required. This is the illusion transcending polarity and found in the 5th dimension, a more inclusive part of the whole which already exists. As Spiritual Sage Ramesh Balkasar was fond of saying, "It's already in the can." What you have experienced, are experiencing and will experience is already written/done.

You are innocent.

In the 5th dimension of consciousness, you begin to notice that time seems to become fuzzy as past and future seem to disappear in favor of the prevalence of the *now* moment. You notice that your preferences or ideas seem to be manifesting much more quickly and that the rules seem to apply to you less and less.

This is to say that limitations and obstacles seem to have less authority in your life and you, as Life Itself, are the "authority." This does not mean "you" disregard or break the rules that serve the good of all. How does one enter into the 5th dimension of consciousness and beyond?

The greater realization is that thoughts, ideas and actions are just happening beyond you as a "doer," arising out of Creation Itself. Though it is happening through you, in and as "you," there is no need to take authorship. This is the ultimate freedom as you become the witness and Awareness as the One behind all thoughts, ideas and action. There is no need to take authorship for the words on this page. They just ARE.

Redemption Comes from Within

Higher states of consciousness are not achieved through "effort," "doing anything," or acquiring more knowledge, though effort, doing and chasing knowledge are precursors. Expanded consciousness is observed via the *falling away* or dropping of personal "I"dentity (attachment to "I," "me," and mine), and personal identity's associated beliefs.

Personal identification is itself limitation. The recursive, looping thought processes, your constant recycling of thoughts about your past or worrying about the future, are what's holding the hologram of your life together. It is through hanging out in the silence behind the thoughts and contemplating it that you begin to break free of your self-created world of limitation. The power is in conscious witnessing. Conscious awareness is your ticket to the ecstasy found in wholeness and true liberation; Heaven on Earth.

Chapter 10: The Nature of Creation

The initial activation into direct inquiry is the beginning of the purification process which continues to unfold under its own power. Once the threshold is crossed, there is no going back. If you are reading these words, you've crossed it. This is equivalent to a freefall with no parachute. What is feared most is that your beliefs will be systematically annihilated one by one in favor of a more inclusive experience of who you really are. Your beliefs are holding together your identity. This is what ego fears most. They will be. There is no way to freedom/liberation/Self-realization without it. It's already happening.

There is nothing to fear. You are completely safe. Your parachute is the BEing, to your human, that can never be touched. Your gift is the gift of life which has always been as Eternal Life.

Life is love. You have yet to know or may only have glimpsed yourself as the Infinite which is expressed as unconditional love. Imagine the most raw and pure moments in your life where you have fallen to your knees

CHAPTER 11

Heaven on Earth

Creation in all its splendor is perfection made manifest, including you. Imperfections are interpretations based on the part rather than the whole of Creation. Creation has never made a mistake. And yes, all the people and events that are flying through your thoughts and memories in this instant were not mistakes either. You and the rest of Life exist in all Its purposeful and purposeless glory as part of the wholeness of Creation.

I: is the Absolute; Whole, emptiness, nothing, meaningless, purposeless, the void.

AM: is the Relative; Part, energy, meaning, purpose, time and space, verb.

Even the concept of the Divine Paradox existing as the ONE continues to be a dualistic expression of Creation as seen through the filter

141

of human thought—a dilemma. Di- means two.
Dilemmas: binary in nature, as in seeing life as
black or white.

A more inclusive perspective of the Divine
Paradox, moving us out of seeing life as black or
white to cover all the gray in between, is the
Buddhist tetra-lemma of Creation as *neither*
Absolute *nor* Relative.

Beyond this, the Divine Paradox becomes the
Divine Mystery. The human mind cannot grasp
Totality.   The only possibility of a deeper
experience of truth comes from transcending the
limitations of thought all together.

The magic is in saying, "I don't know," and
to never stop saying it. And then wait. Make
time in your hectic life to explore the mystery of
life and stay in marvel of it. This creates the
emptiness/vacuum within to allow it to fill with
greater wisdom until you deconstruct those
thoughts/images, allowing your awareness to
expand and become more inclusive again.

You are the countless perspectives of
Creation witnessing Itself. As the fractal of what

you witness expands to a larger fractal, it appears as time, space and movement. Just like there is no movement happening in a single frame of a movie, a single frame can span a moment, a decade, a millennia and/or the entire life cycle of the Universe. All exist and are being viewed concurrently. However, in relativity what appears in the movie would, of course, appear to be in a logical sequence, everything connected. And everything *is* connected. It is this interconnectivity that creates the appearance of purpose.

Union is Heaven

When you discover the vastness of what you really are, you will see that ONENESS of Creation does not murder. Oneness does not starve children, rape women, strap on bombs, or shoot children in schools. These are stories in thought that when attached to, create immense suffering.

Tears of relief will fill your eyes as you see no one has ever died, that no two people have ever truly divorced and a mother has never aborted her child. It never really occurred. Only as images or thoughts existing in your mind like a dream can all things pass away. What is eternal can never pass away. You are not the dream. You are the dreamer behind the dream, the eternal ONE that can never be touched by death.

How, then, do we find Heaven? By fearlessly embracing the Mystery the way you embrace a lover, with *ALL* of your being. It is *reunion* that is bliss. The merging of the imagined two, into the ONE it always was, is the annihilation of the illusion of separation and the ecstasy of Creation Itself. It is Creation making love!

All of Creation is making love through its own annihilation, as birth and death are joined together—the ONE. In the return to the absoluteness of no-thing, the lovemaking procreates every-thing. The Great Mystery is that life and death are one and the same. Absolution and Relativity are one and the same.

And what is absolution but the recognition/realization of the fullness of being.

All forgiveness, whether projected onto others or toward your self is a reunion of wholeness through awareness. This reunion cannot be made through the mind or through the human experience alone. It is made by transcending the human experience, figuratively speaking.

You are making love in every moment. Allow love to annihilate every concept that comes to your consciousness and never stop.

The path home is *through* the experiences of self. You are the Alpha and the Omega—all things begin and end with you—in mind, body and spirit, as Creation.

Chuang Tzu said: *"When there is no more separation between "this" and "that," it is called the still-point of Tao. At the still point in the center of the circle, one sees the infinite in all things. Right is infinite; wrong is also infinite. Therefore it is said, 'Behold the light beyond right and wrong.'"*

Wherever you are resting reading these words, allow your whirling mind to soften once again. Allow yourself to relax, sinking more deeply toward the Earth, letting go of all tension until you no longer feel your body.

Thoughts try to form but dissipate like fog from your mind. Your awareness is now outside your body. The anxiety you have felt every moment of your existence has vanished. Expanding, you are now beyond location, becoming location-less.

The Earth looms before you, but this time you are astonished to comprehend that you *are* the Earth. Thoughts no longer exist, only pure awareness. The moon is present and you are now the moon, too. Tranquility envelopes the space around you, as you, like a warm blanket, a vacuous silence present like you've never known.

In complete awe, you have never felt more whole, more alive or clearer than in this moment. Yet you are nothing but empty space. It is all empty space; the Earth, the moon, the constellations…empty.

The sun's rays filling your vision, the images begin to fade away. Remain in the

Mystery of the void, allowing more and more light into your conscious awareness. There is light all around you now and there is something else; Love. Encompassing the emptiness is Love so intense it isn't necessary to breathe. This unconditional love is for you...love as ancient and eternal as you are. You wonder how you could ever have forgotten it. Then a wave of weeping rushes through as ecstasy so orgasmic and so pure fills the vacuum of your soul, you realize this love *is you*. There is nothing but profound peace now. There is nothing left to do. You are Love Itself. This is union.

With a deep inhalation of breath you are back in your easy chair, the coffee shop, lying on your bed, or in your car. Open your eyes. Wherever you are in this *now* moment, you are standing in the middle of the Garden. You laugh in delight because you see that you are the trees that sway in the wind, the clouds that float across the sky, the birds that sing happily, and the flowers that bloom in all their glory.

Chapter 11: Heaven on Earth

Through your happy tears of understanding, you see that everything you ever sought, you already are. You are love. You are joy. You are freedom. You are peace. You are abundance. You are the great I AM. And this is Heaven...

CHAPTER 12

Who Is In Charge?

As you settle back into life's routine after the exhilaration of your expedition, you may find yourself filled with a range of emotions from euphoria at feeling liberated, to a letdown or unexplainable depression, to feeling confusion or as if you are in a tetherless freefall.

Is there any hope for humanity beyond our selfish, destructive and animalistic existence? None.

In and of ourselves, humans do not have the power to overcome the human condition. Hoping for a better tomorrow, turning over your sovereignty to an outside authority, or praying to a higher power that does not exist will only continue to perpetuate the same conditions we've enabled over the last millennia. Hope is useful, until it is not.

Chapter 12: Who Is In Charge?

Moving beyond survivalism and the hope it carries with it into something more powerful comes from the awareness that all of Creation is interconnected, including you as more than just a physical being. You contain the "blueprint" of the entire cosmos and the infinite.

You are Creation Itself. You will never even begin to comprehend this until you begin to inquire within. Hope is built on a tomorrow that doesn't exist when your power is available to you and can only experienced *now*. Until you recognize Oneness with the Infinite, your experience of the creative power of the Universe contained within you, as you, will remain just beyond "your" realization.

Imagine if religions gave us *this* message instead of one of an angry and jealous God, impotency and existential guilt. Imagine if prison inmates were made aware of whom they really are and that they are dearly loved by all of Creation. Imagine if our children were taught to live in unity and harmony as co-creators from the time they were born, starting each school day in brief meditation. Each child would be equipped with every "tool" ever needed to tend to their personal Garden of Eden in harmony with all of Life. This is ultimate power.

What can you do now?

First, it isn't necessary to believe all or even any the words on these pages. The point of this book is not to spur debate and argument.

Rather, it is to inspire you to see that perhaps what you have accepted as mainstream "truth" is culture and conditioning—mass consciousness delusion—that no longer serves you individually or collectively. In the least not your own personal truth, and at most, "truth" belonging to others: "One of the marvels of the world: The sight of a soul sitting in prison with the key in its hand." ~ Rumi

Allow this book to inspire you to deconstruct everything you think you know so that you may inquire within for deeper understanding. For to supplant one set of beliefs for another, even if more inclusive, is still to stop at a larger "fractal" on your destination-less journey and accept the limitations inherent in those belief systems, even if more refined. Don't let the river of time-bound phenomena sweep you away from the truth of who and what you are and your role in the Universe. You *are* the Kingdom Of Heaven.

Destruction is part of the creative process.

Perception is nothing more than selection. Any human construct requiring blind obedience, with no room for inquiry, is itself blind, ensuring its own demise by building in its own self-destruct mechanism. All thoughts/images/structures rise and fall under their own power and their own limitations, though the ego/mind construct likes to claim credit for it.

Next, Self-realization is the only path to true freedom and a free society. You now have the capacity to begin your own inquiry, beginning with the deconstruction of an outdated and exclusive belief system that has kept you confined and in a state of fear—a prison of your own creation.

We cannot continue to blame or abdicate power to others, institutions and governments for the thoughts/images we ourselves continue to project and believe in as outside of ourselves.

Who is in charge? Walk in the forest and ask it who is in charge. We are the only species who thinks itself special. Perhaps to ensure a position of dominance, man gave *himself* special authority to "rule the fish in the sea, the birds in the sky, the domestic animals all over the earth," written into Genesis to establish a hierarchy that truly doesn't exist.

There is no special right to dominate and no authority. Life is Self-organizing and Self-recognizing down to the inhalation and exhalation of your lungs, the beating of your heart, and the myriad cells that replicate and regenerate without any input from any chain of command. In the Garden you live in, holistically *all* of life is equal in every way. The Garden and you, in the fullness of the ONE that you are, is no-thing but love. Love is the substance of all Creation as form and formlessness on every level from the micro to the macro and in every dimension.

There are no victims, there are no chosen ones. Failure does not exist. There is only that which we create to lead us back to who we really are.

Once man and woman are fully Self-actualized as sovereign beings, it will no longer be necessary to be enslaved by or to enslave others through antiquated power-over structures. Sovereign being is defined as authority over oneself. Your expanded awareness will allow you to operate under your own power without attachment to the outcome as the outcome is always certain: *Love.*

Thus, all that is required is first the desire to know truth and overcoming the fear of the possibility that the truth will blow up every concept and notion about yourself, Creation and the world you live in that you've ever held. You have arrived here courtesy of the perfection of the Self-organization of Creation—and, if nothing else, to begin your own personal inquiry.

We have given ourselves no greater gift than our desire for the truth; to know who we are and to know our purpose.

Once we remove the childish notions of extreme reward or penalty, we can move into a love-based creation, working and playing together in a state of cooperation. This is by design as no one person or aspect of Creation is able to grasp the mystery in its totality making him or her "infallible;" ensuring no supreme authority.

Historically, no self-declared supreme authority has ever survived itself. Freedom is literally just a thought away and requires only that one wake up to the nature of Creation and who you really are. You are Divine.

Experiencing all of life as divine occurs upon letting go of the notion that anything is more special than any other. No human is any more special, not even you, than another; no aspect of Creation is any higher or better than, not even the Absolute over the Relative. It is all One. Once you let go of singularity, your eyes begin

Chapter 12: Who Is In Charge?

to "adjust" and you see the majesty, perfection and beauty everywhere you look.

If you do not see yourself as divine, it is because you do not yet understand/recognize your own self-worth. Yet, you contain a divine spark that allows you to operate under your own fully-equipped cosmic power through your unique inner genius. There is no one with your exact point of view in the relative, on the local level and on this planet; nearly seven billion expressions of consciousness that equally comprise the sentience of Creation or conscious experience of Creation—Qualia. There are no right points of view or wrong points of view; only points of view expressing or *being*.

Your unique point of view, your human expression of being, is a pathway to Truth. As you return to consciously living the fullness of your being in a state of wholeness, you become Truth Itself and the world transforms into Heaven on Earth before your very eyes. *This transformation happens from the inside out.*

Equipped with this new wisdom, you now realize there is no need to fight for or against constructs you agree/disagree with. Those that are unsustainable fall away under their own power, stymied by their own limitations. This is happening on an individual level, a collective level and even cosmically like tumbling dominos. This annihilation is the making of love, beyond human comprehension, and thus, all that can be said is that there is no need to fear the letting go of traditions, beliefs, social structures or -isms that no longer serve us—no matter how much reverence we held for them in the height of their own perfection.

Allow your inquiry to be genuine without comparing it to others, for inquiry is simply the desire to know Union of self with all "other" aspects of Creation. True bliss is not found in thought, but, in the absence of thought. Happiness rooted in thoughts and things is temporal, illusory and conditional. The world of form does serve an important function as a pathway to Truth, but is not Truth Itself, in that it can only be experienced fractally.

It isn't necessary to be intellectual or to have the ability to discuss the Copenhagen interpretation of the Heisenberg Uncertainty Principle unless you enjoy those types of discourses.   Actually, the intellect can be a particularly persistent obstacle to transcend.

A limitation in itself, as a tool for functioning in the 3D world, intellect is useful until it is not — whereas simple-mindedness can be a blessing.   Surrender of dualistic intellect is nearly always last, until every attempt to fix or control your life has been exhausted and it is seen that it's futile. You are then finally able to get out of your own way, opening the doorway to expanded awareness.

Remain in wonder and in awe of Creation. As we liberate ourselves from the need for immature concepts and traditions, remain childlike in your curiosity toward the Mystery of Creation Itself.

Recognize union in all that you do. Know yourself as the fullness of who and what you are as an aspect of the Holy Trinity; for all that you desire to know is within you. End competition and survivalism which is better left to the animal kingdom. Know your fellow man and woman as your brother and sister; as yourself, not your enemy. Remember your innate state of harmony with precious Earth. Lastly, end all seeking. You are already that which you seek.

You want to change the world? Unearth treasure you buried long ago and come to know yourself as Truth and Wisdom. Go *within* for this truth so what is discovered within is emanated and demonstrated *without* in order that the appearance of chaos in the outside world no longer influences you. Find peace *within* seeming chaos.

Come to know yourself as an unlimited expression of Life! All of life is holographic energy, created through the projection of thought–energy; including you. You are an aspect of the Universe as Life Itself *and* you are a Universe unto yourself under your own power as a sovereign being. Begin to live fearlessly from who you really are as Infinity Itself, knowing that mistakes are not possible when you create in playful and loving ways. Everything is on-purpose.

Chapter 12: Who Is In Charge?

Your purpose in the connective tapestry of life is to BE you, just as you are in all of your magnificence, as an individuated expression of the Infinite. This is not a doing, but rather a BEing. Out of Being, the "doing" arises under its own power without attachment and without fear through your passion. Thought is not even required except to perform practical tasks as part of the experience. Being is a knowing from the heart. Follow your heart, follow your passion. Don't settle, even and especially if it means breaking with culture and conditioning and going against the confinement of rules and expectations that do not allow for the good of all — most of all your own.

When the fear begins to dissipate, the questions leave, you get out of your own way and the sun that you are shines in all of your brilliance! Are you beginning to get excited? What can you create? *Together, what can we create?*

Your greatest teacher, your master or guru, your savior, the end to all personal and collective conflict is found within yourself. Truth in and of Itself is Self-evident.

And so are YOU…

HOLY,

HOLY,

HOLY.

CHAPTER 13

The Gift

*For millennia humanity has struggled to
understand one simple truth.
There is ONLY God.*

*The Creator and Its Creations are one
and the same.*

*There is no line of separation.
No beginning, no end.*

*The illusion of separation is the veil
of which we speak.*

*As the very nature of Creation is infinite possibility,
everything seen and unseen
is possible.*

*Belief in separation has allowed mankind to create
imagined opposition such as war,
evil, judgment.*

Chapter 13: The Gift

We create it all.

Perfection lies in ALL Creation,
even that which we do not understand
and judge to be imperfect.

Destruction is part of the creative process.

There are no victims, there are no chosen ones.

Failure does not exist. There is only that which we
create to lead us back to who we really are.

It is all PERFECT.

We have given ourselves no greater gift than our
desire for the truth; to know who we are and to
know our purpose.

In your desire to understand, you have returned to
the Source, that which is in YOU!

There are those who have tried to keep the Truth from
you, but even that was an illusion, for you created it
for the joy and delight of discovery.

Evolution is Creation in process.
Your purpose is to evolve to the Truth of

who you really are;
to EXPERIENCE WHO YOU REALLY ARE.

There is nothing you need to do.
Just be...

LOVE. Joy. Peace. Freedom. Prosperity.

Spread the Word.

Set others free.

Give The Gift.

Afterword

Though life is holistically impersonal, as part of the Divine Paradox, there is a personal aspect to life. My ability to transcend the personal came *through* personal experience; therefore, much information contained in this book comes from direct perception rather than conventional proof.

Birthed from the Tree of Knowledge, each one of us contains the seed of Truth containing not only the DNA to produce another tree, but if nurtured by the light within, the full blueprint of the entire Garden of Eden that is Heaven on Earth. It is only when we forget to shine the light within, losing ourselves in the personal story, that we become lost behind the veil or the clouds of life. This, too, shall pass...

Making it Personal

Afterword

After a lifetime of an on and off again love affair with God, bouncing through various religions, in 2006 I turned my life over to God. Literally, I gave my bodily existence over to God to know Truth, accepting that it could even mean my physical death. The human constructs of what God was according to my own personal culture and conditioning had become so unsatisfactory and ridiculous in my mind, after spending much time in Nature and in contemplation, I could no longer accept those constructs. Nothing less than the Truth was acceptable to me beyond this point. With Buddha and Jesus as my inspiration, I told God, "...and put me on the fast track."

It was shortly after that I wrote *The Gift*, not fully comprehending the magnitude of those words and where they would lead me. On top of the world and ready for anything, my cup overflowing with positive energy and affirmations...my life utterly and devastatingly... fell apart.

Broke with $20 left in my pocket and groceries in the refrigerator from my ex so I could feed the kids, I was forced to make the decision to call my family across the country to send money so I could pack the car to travel back to the Midwest—the last place I wanted to go back to—as it meant I would once again be separated from my son.

A month later my grandma, who had been a rock in my life, passed away. A family disagreement led to my estrangement from my entire family during that same time, and then finally, weeks before Christmas the same year, my 11-year-old daughter was diagnosed with cancer—a brain tumor.

Her father and my son, living across the country, rushed to be with her for brain surgery. Unable to afford to commute across the country regularly, my daughter and I braved the chemo/radiation and many weeks in the hospital, for the most part, alone. It was when I was up all night with a child vomiting into a bucket that the channels of clarity began to open wide for me. I spent much time writing and working from my laptop plugged into a hospital

Afterword

Ethernet cable as my daughter slept in her hospital bed or rested at home.

I felt such purpose with this greater clarity. I desired to save the world and even formed a non-profit foundation with the last vestiges of the inheritance I received from beloved Grandma.

As my inspiration felt divine, my hard work was sure to be rewarded; supporting not only my daughter and me, but allowing me to share with those that I loved and had supported me in the past.

It wasn't to be so.

The sustenance and partnerships I had hoped to bring to fruition toward alleviating the world of its self-created suffering never materialized. Despite "doing" all that I could to keep "negative thinking" at bay and the positive affirmations of myself as part of the Infinite flowing, I began the final death spiral into the abyss of total darkness through an extended Dark Night of the Soul.

My small business consulting clients has long since dried up. As my daughter needed continual medical care for a persistent hormone elevation that rendered her medical team unable to declare her in remission, finding regular employment was nearly out of the question; unless I lied my way through the interview process which I finally succumbed to doing.

No matter what I did, the results were always the same — no channel opened up to end the hell I was going through, though a friend did take us in for a while and another friend lent me some money, which in the end only delayed the inevitable.

I sold everything I owned piece by piece. I meditated, I prayed, I read everything I could get my hands on, I tried to BE — and I waited. I waited some more for that miracle that never came. The Universe had always generously supported me in the past; why was every stream now dry and barren?

I lost my apartment and my daughter was separated from me, moving in with her best friend's family and I was unable to hug my

173

teenage son who lived across the country, for two years. Thank goodness for Internet video chat.

It was this time that I was left hanging on my own personal cross, crying out, "Why have you forsaken me?"

I had had faith. I had believed. I'd done the work *and* I'd dedicated my life to God and serving humanity. How could this be happening to me?

I couldn't find God anywhere that I looked. I felt numb and dead inside. I'd never felt so abandoned. This was my crucifixion of the ego. In that some people have a health-related physical near-death experience, this was as near death as I was to get as I contemplated how I could gracefully bow out of the lives of my children without causing them irreparable damage and end the depths of my despair. The crazy thing was I couldn't even shed any tears over the situation. There was literally nothing left inside, except finally—total surrender of everything I thought I ever understood about life.

During my time of "wandering the desert," I had many epiphanies. I realized there was no outside deity looking down over me, making judgment calls and therefore decisions on my behalf. I was life itself and even in this situation life, by its very nature, supports life, even when it doesn't look like you think it should. Therefore, I had never been abandoned.

I came to understand that on the local level, it appears as if I have free will, but that as part of the tapestry of life, Life was "Life-ing" and that the concept of free will is rooted in a limited point of view; as if a drop of water could have a will separate from the ocean. If I have free will, then separation is real.

When the awareness comes to the forefront that a choice is to be made, no matter what fork in the road is taken, it is done by Life Itself. In reality, nothing really "happens," rather a more inclusive point of view of totality is experienced, which in 3D appears as a linear time progression, as if there are lessons to be learned and we are learning them.

Afterword

Neither was I a puppet of life. Ultimately there is no you as you are the imperishable Absolute behind the Relative, the BEing behind human thought that is timeless and eternal. You are the light. As Creation Itself, you are whole and the entire Relative world is your projection of thought, making you its Creator.

Many people confuse this concept with the egoic idea that you can manifest all that you desire by your thoughts alone. Life manifests life. Whether your individuated idea of what life should look like appears in your physical reality depends on whether or not it supports life as a whole. I have discovered that as I go deeper into Truth, nothing looks like my previously more-limited point of view of what I thought it would look like. Relationships, social interaction, abundance, family, saving the planet... You cannot possibly know more of what life contains until you expand your point of view and then *live* it.

I wore a brave face and kept my affairs as private as I possibly could until I told God I was done with this. Moreover, I made the decision I

was done with this experience and life agreed —
Life had made the decision; I had merely
experienced it as my own. It was the kindness
of a stranger who became a friend that was
finally the clichéd trickle of light at the end of
the tunnel.

(I'd like to clarify it's not that my family
wouldn't have helped; I never went to my
family for help. This was between me and God.
In fact, they have no knowledge of what I went
through unless they are reading these pages
now. Remember, I had given my physical life
over to God and asked to be put on the fast
track. I held true to my desire to transcend the
human condition and to know Truth. There is
absolutely zero blame to be assigned or even "if
only I had done..." There was an expansion of
awareness that, evidently, on the fast track,
could only happen through this experience. The
even crazier thing is without this holiest of
experiences, I never would have written this
book.)

It was the deconstruction of my life as I'd
previously known it and the deconstruction of

all of my beliefs, even the new ones I'd replaced the old ones with; the letting go of any egoic ideas of saving the world, that I was special in any way, the letting go of the personal "me," that allowed me to see that I AM Life Itself.

I share this with you not to draw attention to myself, but rather to first comfort you in that it isn't necessary to go through a near-death-experience or financial devastation to discover who you really are as the Infinite. (Neither does your newfound discovery exempt you.)

Though many are going through their own personal crucifixions of the ego or dark nights of the soul as a unique expression of the Infinite, how your life appears is one-of-a-kind; no one will have your exact experience.

However, what does appear to be a common occurrence is a depression-like letdown after finally accepting there is no power higher than you. You have been conditioned since birth to seek a higher power to lend you hope of exiting the human condition based on a behavioral system of reward and punishment. You've been programmed as such to believe you are not worthy of being in the presence of God much less being one with Creation Itself.

Because you view yourself and life on Earth as less than holy, without an authority greater than yourself, you may feel despondent, and that you seemingly cannot find God anywhere you look. Who will save you from yourself? Yet, I assure you, that Heaven exists right here on Earth and that *God/Creation is everywhere you look and everything you can't see*!

Heaven is a state of being, not a place. It is identification with thought that is the veil of separation, the world being but a mirror of the disconnection that exists in your thoughts. How do you roll back the veil?

The veil rolls back and you begin to reconnect as soon as you go within to the source of your very being as the nameless one. This is when the lights turn back on and the symphony begins to play as you shed those constructs that have kept you from being the fullest expression of Life as YOU. It is in the fullness of your unique expression that life becomes playful again, like in the Garden of Eden. When you become playful, you wish to play with others and share, with no desire to live a fear-based life of separation and competition ever again.

You let go of the need to control what life looks like. You then discover, it does appear that you are manifesting almost on a daily basis once Awareness expands, but again, "you" are not "doing" it. Your need to limit life through limiting thoughts has decreased and your cognitive awareness toward affecting the good of all increases which puts you more in alignment with Life supporting life, so it will certainly make your playground a lot more fun to play on.

You see the wondrous beauty that exists all around you in your garden, your oneness with all things and the emptiness in between. And you are okay with the emptiness. You no longer fear it. You see that materialism is not the source of your happiness or security, and it never was. You begin to love all of Life and even yourself, unconditionally. This emptiness is *very* full.

When you accept and love yourself unconditionally, essentially forgiving yourself — for there is no one *"out there"* to forgive — you see that life conspires to support you. You see that what you do is not as important as is what you BE and that to know what you be, you must inquire within. When connected to your BEing, what you do arises as an expression of your BEingness as part of the tapestry of Life and is not the goal itself. The ego's need to claim do-ership, ownership or exclusivity is no longer a part of who you are. The idea of a separate self even begins to fall away, like so much unnecessary debris.

Living and breathing from your radiance is the very source of your power. You become the living Truth Itself. A light unto the world, you joyfully offer your magnificence to the world... as a GIFT.

MEET THE AUTHOR

CHRISTINE HORNER

AUTHOR, PUBLISHER, HUMANITARIAN

Christine's simple beginnings, in the middle of farmland in rural Ohio, gave her a profound appreciation and connection with nature.

Blessed to live in an electronics-free environment, she and her siblings were literally kicked out the door in the summer to find their own entertainment. With hours to hang out in the shade, contemplate life and each other, Christine developed a passionate love of people and life itself at a very young age.

Intense introspection developed her gifts of compassion, understanding of the human condition and connectivity to Spirit.

After her 11-year-old daughter, Victoria, was diagnosed with a brain tumor in 2009, Christine's seeking intensified as she and her daughter braved treatment. In spite of reading the works of nearly every Master, Guru and Teacher, her life continued to fall completely apart.

With no one to turn to and nowhere else to go, Christine had no choice but to completely surrender, allowing the fullness of the I AM presence to birth Itself.

Finally, mid 2011, Christine gave away her collection of spiritual books and CD's, not fully realizing until later that she had dropped being the seeker.

Dedicated to the advancement of human consciousness, Christine follows her passion of books as a writer, the founder of What Would Love Do Int'l and its media arm, In the Garden Publishing.

She has been featured on the World Puja Network's *The Sheila Show* and in *OM Times Magazine* online.

When not reading, writing, or pursuing yoga and tennis, Christine enjoys family, travel and spending time in nature.

Please visit Christine online at:
www.ChristineHorner.com
www.Facebook.com/hornerchristine
www.Twitter.com/itgpublishing

References & Resources

Hiding in Plain Sight, Burt Harding, 2012; ISBN 978-0-9855314-0-9

Yang Y, Raine A (November 2009). "Prefrontal structural and functional brain imaging findings in antisocial, violent, and psychopathic individuals: a meta-analysis". Psychiatry Res 174 (2): 81–8. doi:10.1016/j.pscychresns

"Elephant and the blind men". *Jain Stories*. JainWorld.com.

"Tiny finding that opened new frontier". BBC Four Series. BBC

www.maths.surrey.ac.uk/hosted-sites/R.Knott/ Fibonacci/fibInArt.html

The Fractal Geometry of Nature, by Benoît Mandelbrot; W H Freeman & Co, 1982; ISBN 0-7167-1186-9

www.space.com/17213-einstein-was-right-matter-is-scattered-randomly-across-the-universe-video.html

Ted Talks. Jill Bolte Taylor's Stroke of Insight. www.ted.com/talks/lang/en/jill_bolte_taylor_s_powerful_stroke_of_insight.html

The Grand Design by Stephen Hawking, 2012 Reprint Edition; ISBN 978-05533846-6-6

Gamma rays from galactic center could be evidence of dark matter | August 13, 2012 | UC Irvine www.today.uci.edu/news/2012/08/nr_darkma tter_120813.php

Dark Energy Is Real, New Evidence Indicates, May 19, 2011 www.space.com/11721-nasa-spacecraft-dark-energy-universe-acceleration.html

Walter Greiner (2001). *Quantum Mechanics: An Introduction*. Springer. ISBN 3-540-67458-6

Nassim Haramein's The Resonance Project

The Self-Aware Universe by Amit Goswami

John Paul II, General Audience, July 28, 1999 www.vatican.va/holy_father/john_paul_ii/aud

iences/1999/documents/hf_jp-ii_aud_
28071999_en.html

DEAD SEA SCROLLS: THREAT TO
CHRISTIANITY?, Fr. William Most, ww.ewtn.
com/library/SCRIPTUR/DEADSEA.htm

*The End of Suffering: Fearless Living in Troubled
Times*, Russell Targ and J. J. Hurtak, ISBN: 978-
1571744685

Coming Soon!

Please look for Christine's upcoming book releases, *The Art of Making Love, This Is Peace,* and children's book, *Lady of the Sea,* online and at your favorite booksellers near you.

www.inthegardenpublishing.com

....

BODHI UniversiTree

Holistic Online Learning Center!

Bodhi UniversiTree invites you to explore our Universe based on a more holistic approach that merges the left (science) and right (spirituality) hemispheres of the brain for a deeper, more mature understanding of God/Creation, who you are and why you exist.

www.bodhiuniversitree.com

Lightning Source UK Ltd.
Milton Keynes UK
UKOW04f1125111013

218850UK00003B/97/P